MULTIHAZARD RISK ATLAS OF MALDIVES

Summary—Volume V

MARCH 2020

© 2020 Asian Development Bank
6 ADB Avenue, Mandaluyong City, 1550 Metro Manila, Philippines
Tel +63 2 8632 4444; Fax +63 2 8636 2444
www.adb.org

Some rights reserved. Published in 2020.

ISBN 978-92-9262-054-7 (print); 978-92-9262-055-4 (electronic); 978-92-9262-056-1 (ebook)
Publication Stock No. TCS200090
DOI: http://dx.doi.org/10.22617/TCS200090

The views expressed in this publication are those of the authors and do not necessarily reflect the views and policies of the Asian Development Bank (ADB) or its Board of Governors or the governments they represent.

ADB does not guarantee the accuracy of the data included in this publication and accepts no responsibility for any consequence of their use. The mention of specific companies or products of manufacturers does not imply that they are endorsed or recommended by ADB in preference to others of a similar nature that are not mentioned.

By making any designation of or reference to a particular territory or geographic area, or by using the term "country" in this document, ADB does not intend to make any judgments as to the legal or other status of any territory or area.

Please contact pubsmarketing@adb.org if you have questions or comments with respect to content, or if you wish to obtain copyright permission for your intended use that does not fall within these terms, or for permission to use the ADB logo.

Corrigenda to ADB publications may be found at http://www.adb.org/publications/corrigenda.

Notes:
In this publication, "$" refers to United States dollars.
The maps presented in this atlas reflect airports based on 2017 data from the Civil Aviation Authority of Maldives.

On the cover: An aerial view shows 1 of 26 natural atolls that make up Maldives, which also includes nearly 1,200 small coral islands and some of the world's most beautiful beaches. Recognized as the seventh-largest in the world, the coral reefs and associated ecosystems of Maldives are key foundations for food security and means of livelihood. Yet, they are considered as among the most vulnerable to climate change (photo by Roberta Gerpacio).

Contents

Tables and Maps   vi

Foreword   viii

Acknowledgments   ix

Abbreviations   x

Risk Mapping: Making the Invisible Visible   1

Paradise at Risk   3

Geography of Maldives   4
    Exclamation Point on the Ocean   4

Dynamic Geography   6

Land Management   10

Dynamic Climate   13

Automatic Weather Stations and Meteorological   14
Observation Stations

Historical Annual Climate   16
    Rainfall   16
    Temperature   16

Historical Seasonal Climate   19
    Seasonal Average Rainfall (1970–2005)   20
    Seasonal Average Temperature (1970–2005)   21

Future Climate   **22**

    Annual Average Rainfall (RCP 4.5)   23

    Average Seasonal Rainfall Projection (DJF, RCP 4.5)   24

    Average Seasonal Rainfall Projection (MAM, RCP 4.5)   25

    Average Seasonal Rainfall Projection (JJA, RCP 4.5)   26

    Average Seasonal Rainfall Projection (SON, RCP 4.5)   27

    Average Annual Rainfall Projection (RCP 8.5)   28

    Average Seasonal Rainfall Projection (DJF, RCP 8.5)   29

    Average Seasonal Rainfall Projection (MAM, RCP 8.5)   30

    Average Seasonal Rainfall Projection (JJA, RCP 8.5)   31

    Average Seasonal Rainfall Projection (SON, RCP 8.5)   32

    Average Annual Temperature Projection (RCP 4.5)   33

    Average Seasonal Temperature Projection (DJF, RCP 4.5)   34

    Average Seasonal Temperature Projection (MAM, RCP 4.5)   35

    Average Seasonal Temperature Projection (JJA, RCP 4.5)   36

    Average Seasonal Temperature Projection (SON, RCP 4.5)   37

    Average Annual Temperature Projection (RCP 8.5)   38

    Average Seasonal Temperature Projection (DJF, RCP 8.5)   39

    Average Seasonal Temperature Projection (MAM, RCP 8.5)   40

    Average Seasonal Temperature Projection (JJA, RCP 8.5)   41

    Average Seasonal Temperature Projection (SON, RCP 8.5)   42

Summary of Observations for Rainfall   **43**

Summary of Observations for Temperature   **44**

Geophysical Hazards   **45**

Demographics and Economy   **50**

Population   **51**

Education   **54**

Health   **56**

Lives Centered at Sea   **58**

Tourism   60

Transportation   62

Harbor Facilities   65

Sand Mining Applications   67

Power Stations   70

Human Development Index   71

Environmentally Sensitive Areas   73

Coastal Protection   75

Marine Conservation and Biodiversity   78

Threats to Marine and Coastal Biodiversity   80
    Coastal Erosion   80
    Coral Bleaching   81

Disaster Risk in Maldives   83

Map Data Sources   91

References   93

Tables and Maps

Tables

V.1	Islands with Reclamation	6
V.2	Land Use and Land Cover, Maldives	10
V.3	Islands with Weather Stations	14
V.4	Variation in Rainfall Patterns across Space, Time, and Representative Concentration Pathways	43
V.5	Variation in Temperature Patterns across Space, Time, and Representative Concentration Pathways	44
V.6	Atoll Population in Maldives	51
V.7	Number of Schools and Student Capacity per Atoll, Maldives	55
V.8	Types of Hospitals in Maldives	56
V.9	Number of Hospitals per Atoll, Maldives	56
V.10	Number of Resorts per Atoll, Maldives	60
V.11	Airports in Maldives	63
V.12	Harbors in Maldives	65
V.13	Sand Mining in Maldives	67
V.14	Maldives, Islands with Coastal Protection	75
V.15	Maldives, Islands with Coral Reef Monitoring Sites	78
V.16	Hazard Categories and Index Ranges	83

Maps

V.1	Maldives Atolls and Islands	4
V.2	Maldives, Basemap	5
V.3	Maldives, Land Reclamation	9
V.4	Maldives, Meteorological Observation Stations	15
V.5	Maldives, Annual Average Rainfall (1970-2005)	17
V.6	Maldives, Annual Average Temperature (1970-2005)	18
V.7	Maldives, Seasonal Average Rainfall (1970–2005)	20
V.8	Maldives, Seasonal Average Temperature (1970–2005)	21
V.9	Maldives, Annual Average Rainfall Projection (RCP 4.5)	23
V.10	Maldives, Average Seasonal Rainfall Projection (DJF, RCP 4.5)	24
V.11	Maldives, Average Seasonal Rainfall Projection (MAM, RCP 4.5)	25

V.12	Maldives, Average Seasonal Rainfall Projection (JJA, RCP 4.5)	26
V.13	Maldives, Average Seasonal Rainfall Projection (SON, RCP 4.5)	27
V.14	Maldives, Average Annual Rainfall Projection (RCP 8.5)	28
V.15	Maldives, Average Seasonal Rainfall Projection (DJF, RCP 8.5)	29
V.16	Maldives, Average Seasonal Rainfall Projection (MAM, RCP 8.5)	30
V.17	Maldives, Average Seasonal Rainfall Projection (JJA, RCP 8.5)	31
V.18	Maldives, Average Seasonal Rainfall Projection (SON, RCP 8.5)	32
V.19	Maldives, Average Annual Temperature Projection (RCP 4.5)	33
V.20	Maldives, Average Seasonal Temperature Projection (DJF, RCP 4.5)	34
V.21	Maldives, Average Seasonal Temperature Projection (MAM, RCP 4.5)	35
V.22	Maldives, Average Seasonal Temperature Projection (JJA, RCP 4.5)	36
V.23	Maldives, Average Seasonal Temperature Projection (SON, RCP 4.5)	37
V.24	Maldives, Average Annual Temperature Projection (RCP 8.5)	38
V.25	Maldives, Average Seasonal Temperature Projection (DJF, RCP 8.5)	39
V.26	Maldives, Average Seasonal Temperature Projection (MAM, RCP 8.5)	40
V.27	Maldives, Average Seasonal Temperature Projection (JJA, RCP 8.5)	41
V.28	Maldives, Average Seasonal Temperature Projection (SON, RCP 8.5)	42
V.29	Maldives, Cyclonic Wind Hazard Zone	46
V.30	Maldives, Surge Hazard Zone	47
V.31	Maldives, Seismic Hazard Zone	48
V.32	Maldives, Tsunami Hazard Zone	49
V.33	Maldives, Population	52
V.34	Maldives, Population Density	53
V.35	Maldives, Healthcare Facilities	57
V.36	Maldives, Resort Islands	61
V.37	Maldives, Transportation	64
V.38	Maldives, Harbor Facilities	66
V.39	Maldives, Approved Sand Mining Locations	69
V.40	Maldives, Power Stations	70
V.41	Maldives, Human Development Index	72
V.42	Maldives, Environmentally Protected and Sensitive Areas	74
V.43	Maldives, Coastal Protection	77
V.44	Maldives, Coral Bleaching Risk Assessment	82
V.45	Maldives, Rain-Induced Flood (Islands)	84
V.46	Maldives, *Udha* Hazard (Islands)	85
V.47	Maldives, Wind and Wave Hazards	86
V.48	Maldives, Tsunami Hazard Rank (Islands)	87
V.49	Maldives, Hydrometeorological Multihazard (Islands)	88
V.50	Maldives, Multihazard Physical Risk Index	89
V.51	Maldives, Multihazard Social Risk Index	90

Foreword

Maldives is among the countries most vulnerable to the impacts of climate change as it is a small island nation with extremely low elevations. Maldives is also very vulnerable to impacts of rising air and sea surface temperatures and changes in rainfall patterns. Climate change impacts will therefore impose significant negative consequences on the Maldivian economy and society. Some of the priority vulnerabilities to climate change are land loss and beach erosion, infrastructure damage, degradation of coral reefs, and adverse impacts on water resources, food security, human health, and the overall economy.

Sustainable coastal resources management is of particular importance to Maldives, such that all regulations involving various development activities have coastal components. Despite the government's continued efforts in improving and sustaining coastal resources management, critical issues remain, such as the need for systematized coastal monitoring, clear definition of coastal boundaries and coastal development, enhanced regulatory and monitoring capacities for coastal resources protection, and sustainable long-term strategies on land reclamation and marine area protection. At a time when climate is rapidly changing and extreme weather events are frequently occurring, the critical roles that marine and coastal environments play in mitigating and adapting to climate change need to be sufficiently documented and properly recognized. It is therefore essential for Maldives to develop and establish a comprehensive digital database of marine and coastal ecosystem features and services that can be regularly monitored.

The *Multihazard Risk Atlas of Maldives* was developed through the project "Establishing a National Geospatial Database for Mainstreaming Climate Change Adaptation into Development Activities and Policies in Maldives" under the Asian Development Bank's regional knowledge and support (capacity development) technical assistance Action on Climate Change in South Asia (2013–2018). This five-volume atlas aims to promote the sustainable development of coastal and marine ecosystems and their various components, by enhancing the awareness of stakeholders on and enjoining them to address climate and disaster risks (including hazards, exposures, and vulnerabilities) to which ecosystems are exposed. The atlas presents spatial information and maps necessary for assessing future development investments in terms of their risks to climate and geophysical hazards.

The target audience of the *Multihazard Risk Atlas of Maldives* are the concerned stakeholders with current or planned development activities in the country, including public and private sectors, nongovernment organizations, research and academic community, development partner agencies, other financial institutions, and the general public. The atlas will also be a useful reference for other developing countries with similar geographical and environmental conditions, particularly small island developing states. It is envisioned that the atlas will significantly contribute to rendering important sector development investments more resilient to hazard-specific risk scenarios in the short, medium, and long terms.

H.E. Dr. Hussain Rasheed Hassan
Minister
Ministry of Environment, Malé

Shixin Chen
Vice-President for Operations 1
Asian Development Bank, Manila

Acknowledgments

Government Ministries, Departments, and Agencies in Maldives
   Civil Aviation Authority
   Land and Survey Authority
   Marine Research Institute
   Meteorological Service
   Ministry of Economic Development
   Ministry of Education
   Ministry of Environment
   Ministry of Fisheries, Marine Resources and Agriculture
   Ministry of Health
   Ministry of National Planning and Infrastructure
   Ministry of Tourism
   National Bureau of Statistics
   National Disaster Management Center

International Institutions
   Manila Observatory
   Marine Spatial Ecology Lab, University of Queensland, Australia
   SANDER + PARTNER
   United Nations Development Programme

International Institutions in Maldives
   International Union for Conservation of Nature, Maldives
   United Nations Development Programme, Maldives

National Consultant Team
   Ahmed Jameel, Integrated Coastal Zone Management Specialist
   Faruhath Jameel, Geographic Information Systems Specialist and Team Leader
   Hussain Naeem, Coastal Ecosystems and Biodiversity Specialist
   Mahmood Riyaz, Climate Change Risk Assessment Specialist

Abbreviations

°C	–	Celsius
AWS	–	automated weather station
BODC	–	British Oceanographic Data Centre
CAA	–	Maldives Civil Aviation Authority
DJF	–	December, January, February
EPA	–	Maldives Environmental Protection Agency
GEBCO	–	General Bathymetric Chart of the Oceans
GHCN	–	Global Historical Climatology Network
ha	–	hectare
IHO	–	International Hydrographic Organization
IOC	–	Intergovernmental Oceanographic Commission
IUCN	–	International Union for Conservation of Nature
JJA	–	June, July, August
km^2	–	square kilometer
kW	–	kilowatt
MAM	–	March, April, May
ME	–	Ministry of Environment
MED	–	Ministry of Economic Development
MFMRA	–	Ministry of Fisheries, Marine Resources and Agriculture
MLSA	–	Maldives Land and Survey Authority
mm/day	–	millimeter per day
MMS	–	Maldives Meteorological Service
MNPI	–	Ministry of National Planning and Infrastructure
MOE	–	Ministry of Education
MOH	–	Ministry of Health
MOT	–	Ministry of Tourism
MRC	–	Marine Research Centre
NBS	–	National Bureau of Statistics
RCP	–	representative concentration pathway
SON	–	September, October, November
UNDP	–	United Nations Development Programme
UTM	–	Universal Transverse Mercator
WGS	–	World Geodetic System

Risk Mapping: Making the Invisible Visible

1

Maps have changed the way we see the world. By symbolizing the features of the earth and drawing its visible and invisible boundaries, maps give humans a wider perspective, which allows us to understand the patterns, trends, and interconnected components of our planet and see beyond where we have traveled. What used to be invisible to some became visible for all through maps.

At the global scale, the size of Maldives fades in comparison with its neighboring countries such as India and Sri Lanka. Sitting in the middle of the Indian Ocean, Maldives is barely visible but greatly vulnerable to natural hazardous elements, and it strives to overcome these challenges. People have lived in this beautiful nation for thousands of years despite limited land space and multiple hazards. However, their lives, livelihoods, and properties are becoming increasingly exposed to hydrometeorological, climatological, and geological hazards.

The *Multihazard Risk Atlas of Maldives* examines the often-unexplored networks of human and environmental paradigms, producing a compound picture of risk that hinders development. It compiles the maps generated for Action on Climate Change in South Asia: Establishing a National Geospatial Database for Mainstreaming Climate Change Adaptation into Development Activities and Policies in Maldives, an Asian Development Bank–Maldives Ministry of Environment technical assistance (TA) project.

The national and atoll-scale maps were produced to paint a clearer picture of climate and disaster risk in Maldives and can hopefully be useful in formulating action plans and policies and in communicating risk information toward averting economic losses and harm. They can become useful tools in visually comparing spatial and temporal variations, spotting interaction among layers, and in guiding possible development directions considering future climate scenarios in the context of the changing geography, demographics, climate, economy, and environmental conditions of Maldives.

This atlas is divided into thematic sections: geography, climate and geophysical hazards, economy and demographics, and biodiversity. A summary volume provides the highlights and key messages from the overall TA activities in Maldives. The main goal of this atlas is to capture often-unmapped factors shaping Maldives' contrasting picture of paradise in the midst of an evolving world, which generates natural hazards to the islands.

Map of the world. A map visualizing the different countries of the world (photo from Crates, World Map without Boundaries, Creative Commons).

Paradise at Risk

In the middle of the Indian Ocean, southwest of India, lies a chain of atolls with islands forming a small low-lying island nation called Maldives. These tropical islands, emerging from corals and sprinkled across turquoise crystal calm waters caressing white sand beaches, paint a picture of a paradise.

The flatlands and surrounding seas form mangrove and aquatic ecosystems supporting the life of the communities that inhabit a tenth of the nation's total islands. These communities have witnessed the continuous formation and transformation of the islands as a product of natural tidal action eroding and depositing sand or of human hands reclaiming or dredging the land. They are also highly vulnerable to strong waves during storms or earthquakes (Waheed and Shakoor 2015). Moreover, a 1-meter high increase in sea level due to a warmer world (Khan et al. 2002) could inundate two-thirds of the country's land (Ahmed and Suphachalasai 2014). Rising seas could wipe away Maldives as most of its islands are less than 3 meters high (Khan et al. 2002). In addition, the natural biota in the islands might not be able to cope with the fast-paced development and could be displaced (ADB 2017a). Geologic events and the changing climate pose a threat to the future of the islands and its inhabitants.

This volume summarizes the contents of the other four volumes of the *Multihazard Risk Atlas of Maldives*.

Golden sunset. The usual serene and calm afternoon in a shore of Maldives. This golden reflection of the sun in the crystal clear Maldives' water makes it a good time to sit, relax, and see the beauty of nature (photo by P.K. Niyogi).

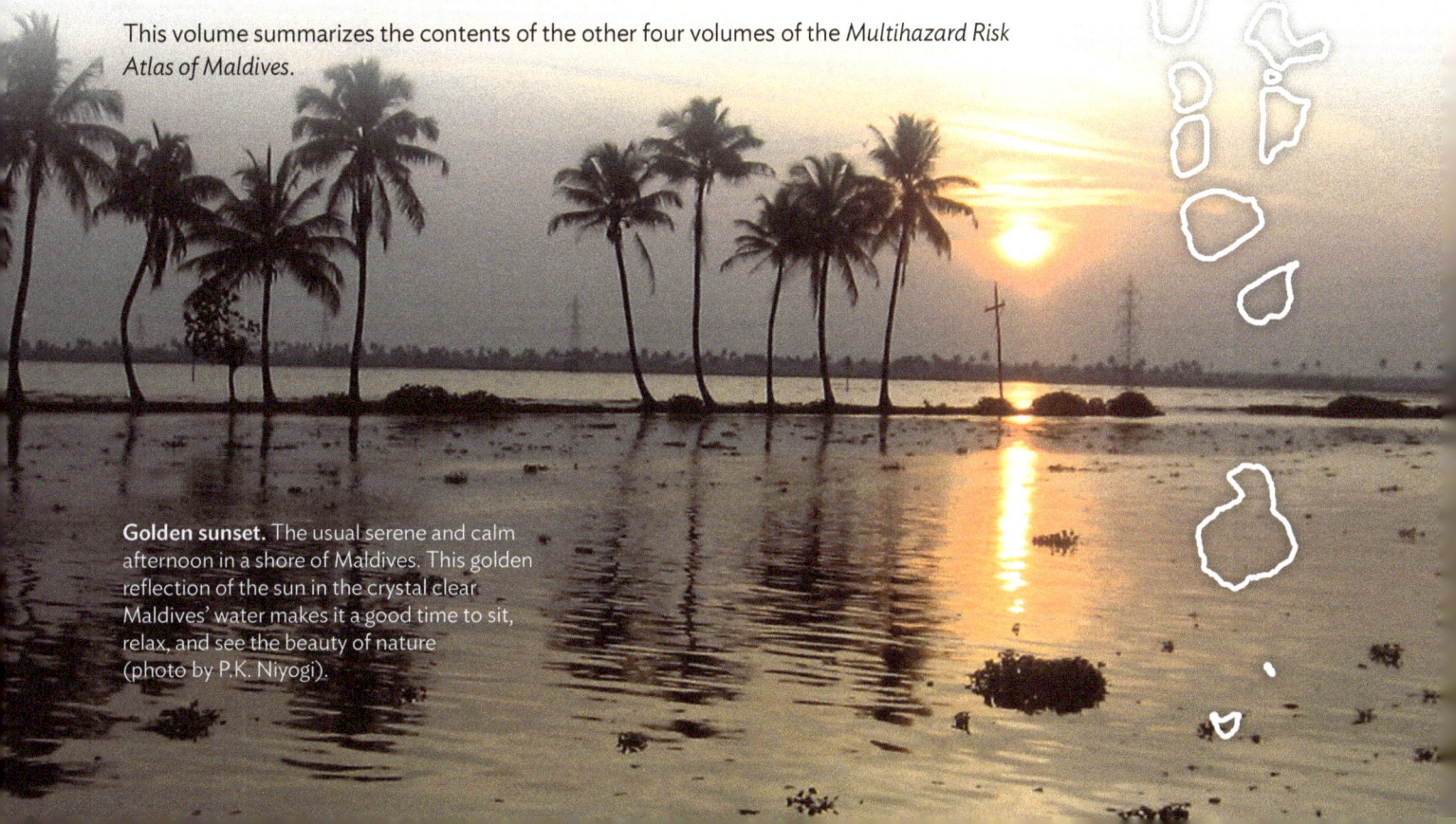

Geography of Maldives

Exclamation Point on the Ocean

Maldives sits like an exclamation point, marking the middle of the Indian Ocean and drawn vertically almost at the equator. Currently, there are about 1,900 islands forming 26 atolls scasttered across 870 kilometers from north to south. Less than 200 islands are inhabited. These numbers might change in the coming years due to the dynamic shifts in factors shaping the island such as sea level and land reclamation. Islands rising up to 2.4 meters above mean sea level subtly punctuate the flatness of the sea.

This summary explores the atolls, transformation of the islands through land reclamation, and how the land is being managed.

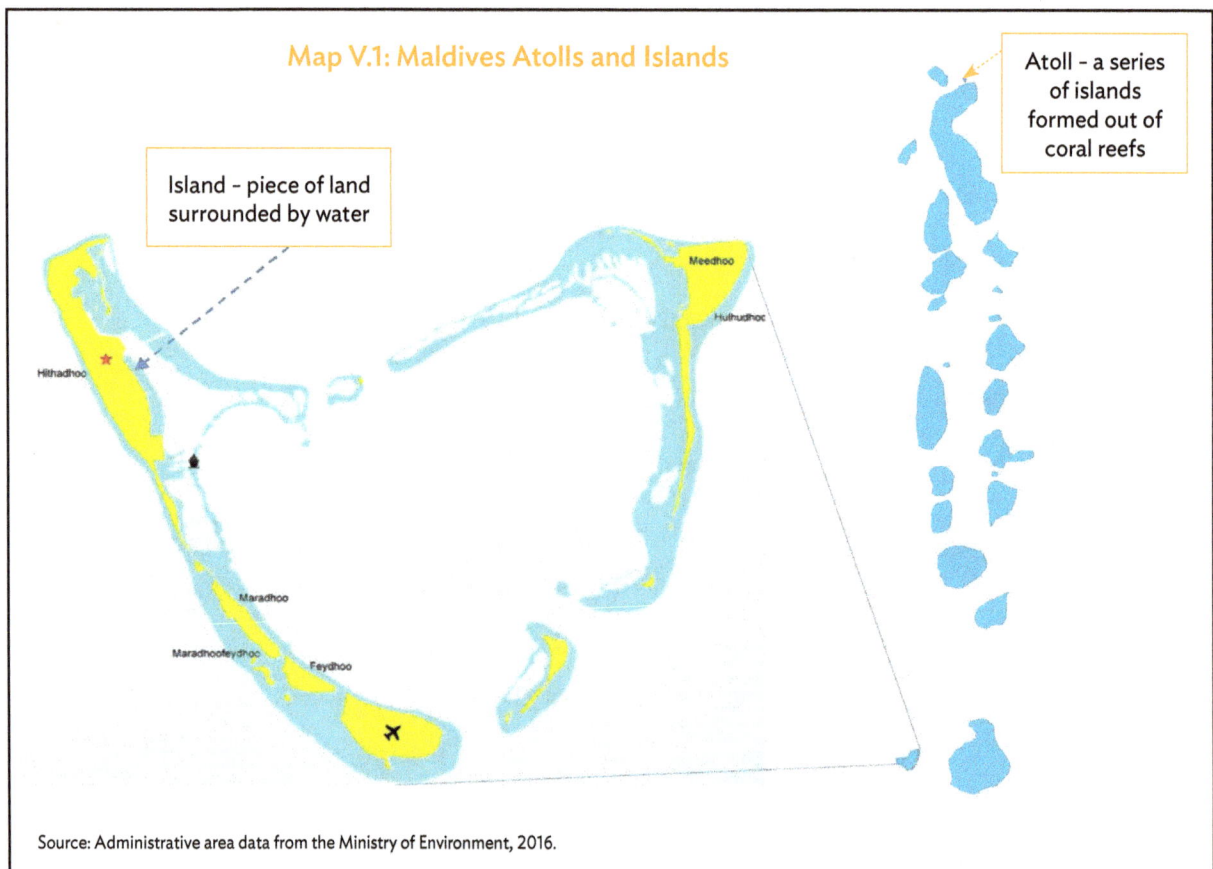

Map V.1: Maldives Atolls and Islands

Island - piece of land surrounded by water

Atoll - a series of islands formed out of coral reefs

Meedhoo

Hulhudhoo

Hithadhoo

Maradhoo

Maradhoofeydhoo

Feydhoo

Source: Administrative area data from the Ministry of Environment, 2016.

Map V.2: Maldives, Basemap

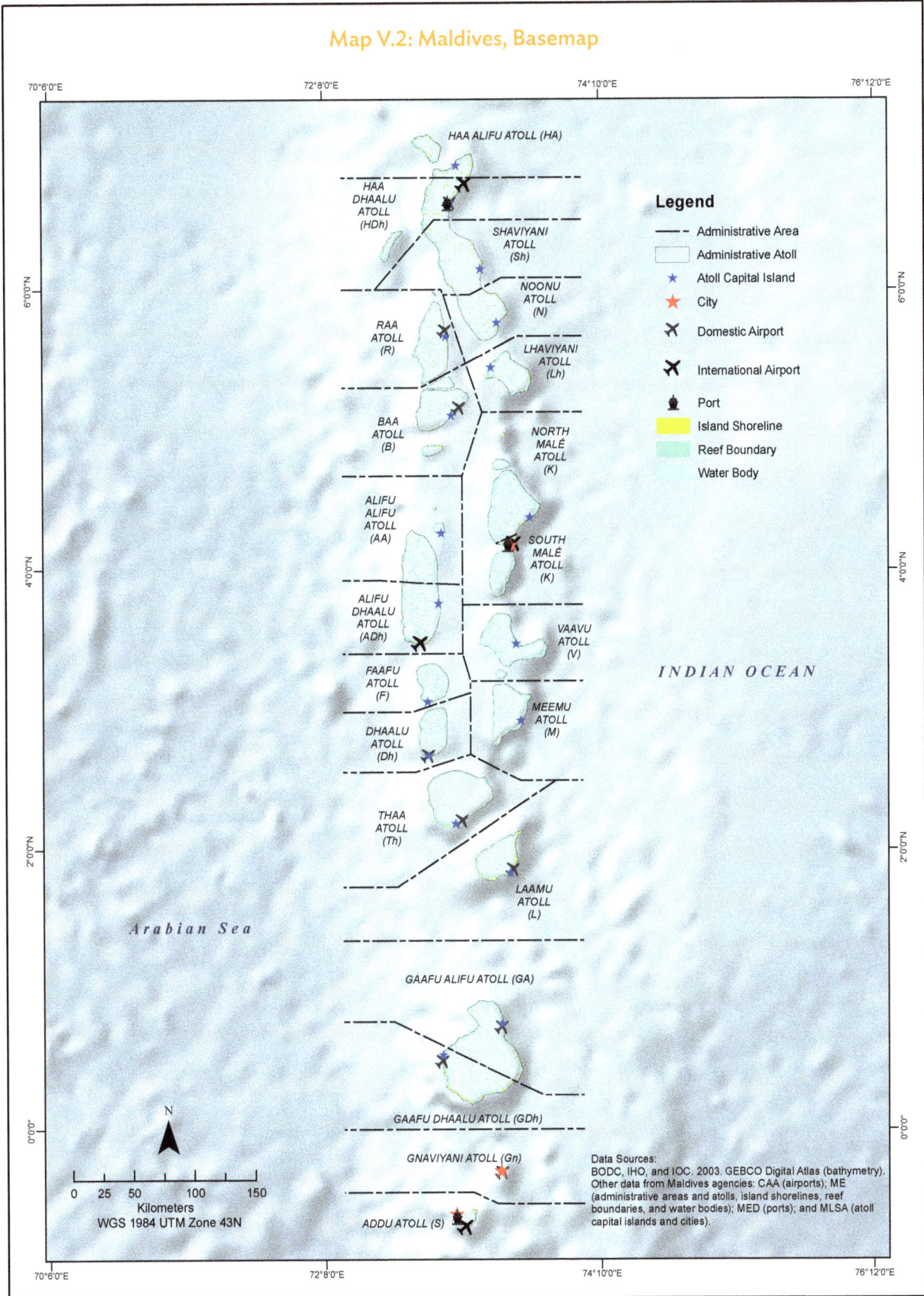

Legend

- – – – Administrative Area
- ☐ Administrative Atoll
- ✦ Atoll Capital Island
- ✦ City
- ✈ Domestic Airport
- ✈ International Airport
- ⚓ Port
- ▮ Island Shoreline
- ▮ Reef Boundary
- Water Body

HAA ALIFU ATOLL (HA)
HAA DHAALU ATOLL (HDh)
SHAVIYANI ATOLL (Sh)
NOONU ATOLL (N)
RAA ATOLL (R)
LHAVIYANI ATOLL (Lh)
BAA ATOLL (B)
NORTH MALÉ ATOLL (K)
ALIFU ALIFU ATOLL (AA)
SOUTH MALÉ ATOLL (K)
ALIFU DHAALU ATOLL (ADh)
VAAVU ATOLL (V)
FAAFU ATOLL (F)
MEEMU ATOLL (M)
DHAALU ATOLL (Dh)
THAA ATOLL (Th)
LAAMU ATOLL (L)
GAAFU ALIFU ATOLL (GA)
GAAFU DHAALU ATOLL (GDh)
GNAVIYANI ATOLL (Gn)
ADDU ATOLL (S)

INDIAN OCEAN

Arabian Sea

N

Kilometers
0 25 50 100 150
WGS 1984 UTM Zone 43N

Data Sources:
BODC, IHO, and IOC. 2003. GEBCO Digital Atlas (bathymetry). Other data from Maldives agencies: CAA (airports); ME (administrative areas and atolls, island shorelines, reef boundaries, and water bodies); MED (ports); and MLSA (atoll capital islands and cities).

Dynamic Geography

The natural geographic formation of Maldives presents challenges in urban expansion and development. Limited dry land combined with urban growth requires measures to maximize the available space for multiple activities. One mode of urban expansion is land reclamation. Land reclamation has become part of the development in the country and a solution to the limited available land, transforming the islands over the years to accommodate more human activities. Land reclamation is noticeable in Thinadhoo Island located in Gaafu Dhaalu Atoll. The map of Thinadhoo Island shows straightened coasts, which indicate land reclamation. Other locations of land reclamation in Maldives are listed in Table V.1 and shown in Map V.3.

Table V.1: Islands with Reclamation

Atoll	No. of Islands with Reclamation	Islands with Reclamation
Haa Alifu	1	Dhidhoo
Haa Dhaalu	1	Kulhudhuffushi
Addu City	2	Hithadhoo
		Feydhoo
Alifu Alifu	2	Bodufolhudhoo
		Rasdhoo
Alifu Dhaalu	2	Dhihdhoo
		Maamigili
Faafu	2	Magoodhoo
		Nilandhoo
Laamu	2	Gaadhoo
		Kunahandhoo
Lhaviyani	2	Hinnavaru
		Naifaru
Noonu	2	Maafaru
		Vavathi
South Malé	2	Gulhi
		Maafushi

continued on next page

Table V.1 *continued*

Atoll	No. of Islands with Reclamation	Islands with Reclamation
Dhaalu	3	Kudahuvadhoo
		Maaen'boodhoo
		Meedhoo
Gaafu Alifu	3	Kolamaafushi
		Falhuverrahaa
		Dhaandhoo
Baa	4	Dharavandhoo
		Eydhafushi
		Thulhaadhoo
		Fares
Gaafu Dhaalu	4	Hoan'dedhdhoo
		Thinadhoo
		Faresmaathodaa
		Rodhavarrehaa
Meemu	4	Dhiggaru
		Maduhvari
		Muli
		Naalaafushi
Shaviyani	4	Milandhoo
		Funadhoo
		Komandoo
		Maroshi
Thaa	6	Vilufushi
		Madifushi
		Dhiyamigili
		Guraidhoo
		Hirilandhoo
		Thimarafushi
North Malé	8	Gaafaru
		Thulusdhoo
		Himmafushi
		Hulhumalé
		Malé
		Vilin'gili
		Hulhulé
		Tilafushi

Source: Government of Maldives, Ministry of National Planning and Infrastructure, 2017.

Land reclamation. To accommodate more people and urban activities, the coasts of major islands have been reclaimed. The Government of Maldives prioritized the Hulhumalé Land Reclamation and Development Project to address the need for more space around Malé City. This project will expand the island by 12.8 square kilometers in a sustainable manner (ADB 2017a) (photo by Shahee Ilyas).

Map V.3: Maldives, Land Reclamation

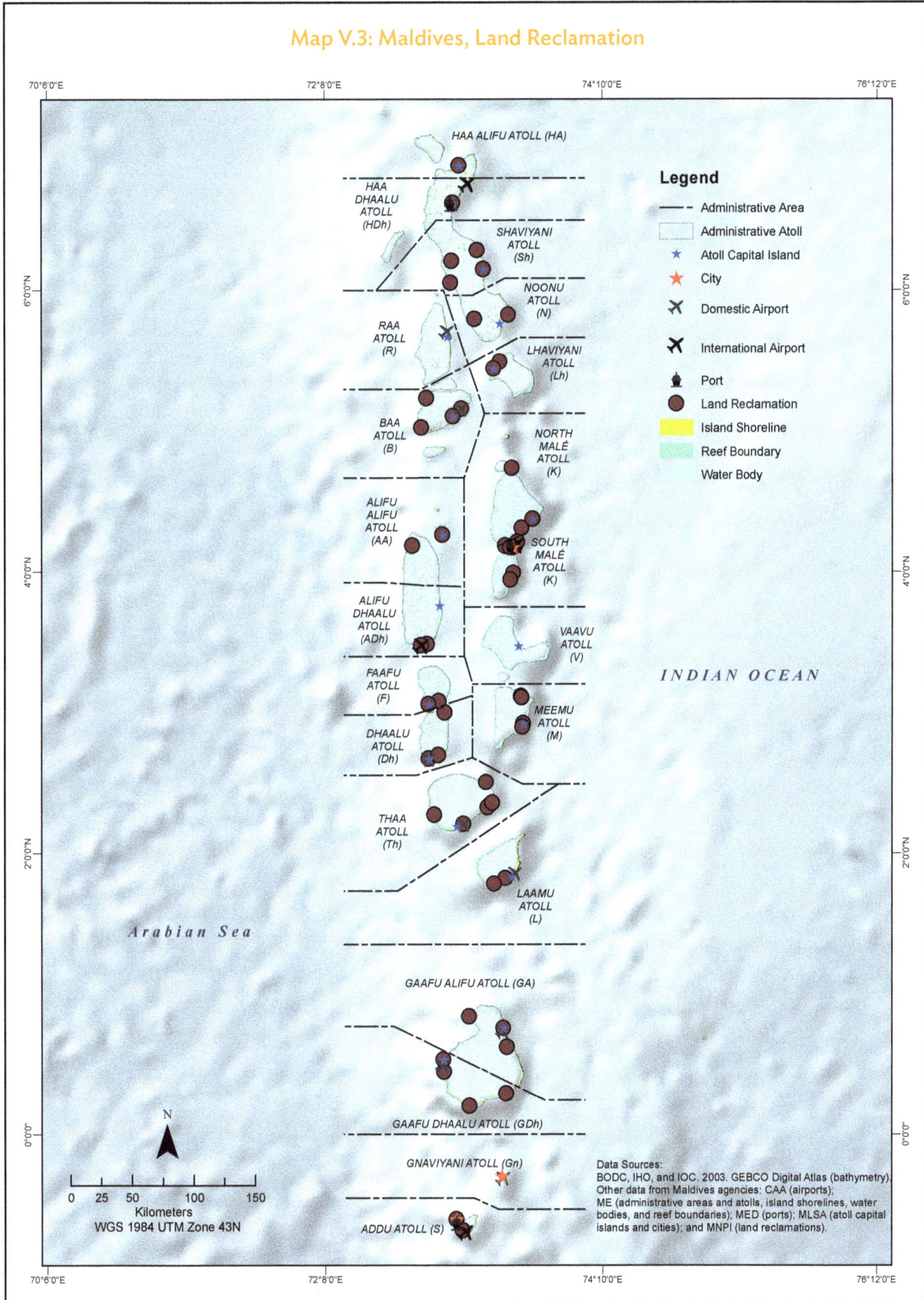

HAA ALIFU ATOLL (HA)

HAA DHAALU ATOLL (HDh)

SHAVIYANI ATOLL (Sh)

NOONU ATOLL (N)

RAA ATOLL (R)

LHAVIYANI ATOLL (Lh)

BAA ATOLL (B)

NORTH MALÉ ATOLL (K)

ALIFU ALIFU ATOLL (AA)

SOUTH MALÉ ATOLL (K)

ALIFU DHAALU ATOLL (ADh)

VAAVU ATOLL (V)

FAAFU ATOLL (F)

MEEMU ATOLL (M)

DHAALU ATOLL (Dh)

THAA ATOLL (Th)

LAAMU ATOLL (L)

GAAFU ALIFU ATOLL (GA)

GAAFU DHAALU ATOLL (GDh)

GNAVIYANI ATOLL (Gn)

ADDU ATOLL (S)

INDIAN OCEAN

Arabian Sea

Legend

- — — Administrative Area
- ☐ Administrative Atoll
- ★ Atoll Capital Island
- ★ City
- ✕ Domestic Airport
- ✈ International Airport
- ⚓ Port
- ⬤ Land Reclamation
- ▮ Island Shoreline
- ▮ Reef Boundary
- Water Body

N

0 25 50 100 150
Kilometers
WGS 1984 UTM Zone 43N

Data Sources:
BODC, IHO, and IOC. 2003. GEBCO Digital Atlas (bathymetry).
Other data from Maldives agencies: CAA (airports);
ME (administrative areas and atolls, island shorelines, water
bodies, and reef boundaries); MED (ports); MLSA (atoll capital
islands and cities); and MNPI (land reclamations).

70°6'0"E   72°8'0"E   74°10'0"E   76°12'0"E

6°0'0"N

4°0'0"N

2°0'0"N

0°0'0"

Land Management

Land space is a necessary resource for development. Maldives has only less than 250 square kilometers (km²) of land. With its limited available land, allocation of space for specific land use is a challenge. Currently, the country has more vegetated land cover (shrubs, herbs, forest, and palm trees) than built-up areas (high density urban areas, road, airport, and low density urban areas). Huge land is allocated for beaches (20.9 km²) and island resorts (16.8 km²). A small portion (6.7 km²) is for agriculture. Other spaces are classified as inland water and wetlands (Table V.2).

Table V.2: Land Use and Land Cover, Maldives

Land Use and Land Cover	Area (km²)	Percentage
Land Use		
High density urban areas	1.79	0.76
Roads	4.66	1.98
Airports	5.80	2.46
Harbors	7.32	3.11
Island resorts	16.78	7.12
Low density urban areas	29.62	12.56
Agricultural areas	6.65	2.82
Land Cover		
Inland waters	2.97	1.26
Wetlands	3.39	1.44
Barren, sparsely vegetated areas	8.88	3.77
Beaches and sand	20.85	8.84
Shrubs and/or herbaceous vegetation areas	36.07	15.30
Forests	40.86	17.33
Palm trees	50.10	21.25
Total	**235.74**	**100.00**

km² = square kilometer.

Note: Total may not add up due to rounding.

Source: Government of Maldives, Ministry of Environment, 2016.

Land cover. Maldives' national tree, the palm tree, covers a fifth of the country's land area. Together with palm trees, forests and shrublands cover more than half of the country's land (photo by Shifaaz Shamoon).

Buildings stand next to one another along the narrow roads of Malé City in Maldives. Less than 1% of Maldives has high density urban land use classification (photo by Shahee Ilyas).

Dynamic Climate

Maldivians enjoy mostly sunny days. Warm temperature (ranging from 26°C to 33°C) is experienced throughout the year, with little variation, for the entire stretch of Maldives (Moosa 2014).

While temperature stays relatively constant, the rainfall varies depending on the monsoon. Maldives has two monsoon seasons: (i) northeast monsoon or dry season, starting in November and lasting until April; and (ii) southwest monsoon, which brings rainfall from May to October. Rainfall varies across the year and latitude (Moosa 2014).

In Maldives, climate is an important element shaping present and future policies, actions, and decisions. It affects Maldives' coastal and marine resources (Asian Development Bank 2015). Increased global temperature can raise sea levels and inundate the low-lying islands of the country (Khan et al. 2002). Furthermore, the corals would also be damaged by an increase in global temperatures. Water resources as well as agriculture also rely on sufficient precipitation to meet the water and food requirements of the communities. Natural flora and fauna require moderate climate to survive.

To prevent losses due to weather- and climate-related disaster events, it is important to monitor the weather parameters and evaluate the climatic trends in the country. This is done through data gathered from the automatic weather stations and meteorological observation stations scattered in different islands.

Rainbow in horizon. A beautiful rainbow visible in the horizon of a beach in Maldives. Rainbows are formed when there is a refraction and reflection of light in cloud droplets, forming a spectrum of light (photo by Michael Pfütze).

Automatic Weather Stations and Meteorological Observation Stations

Maldives has a total of 18 weather monitoring stations, 5 of which are meteorological observation stations (Table V.3). The locations of the weather monitoring stations are concentrated in the southern region, particularly in Addu and Gaafu Dhaalu atolls. Most of the meteorological observation stations are located in the capital islands and international airports (Map V.4).

Table V.3: Islands with Weather Stations

ATOLL	Island	Station
Addu City	Hithadhoo	Automatic Weather Station
Addu City	Gan	Meteorological Observation Station
Addu City	Meedhoo	Automatic Weather Station
Alifu Alifu	Rasdhoo	Automatic Weather Station
Alifu Dhaalu	Rangali	Automatic Weather Station
Gaafu Alifu	Villingili	Automatic Weather Station
Gaafu Dhaalu	Kaadedhdhoo	Meteorological Observation Station
Gaafu Dhaalu	Gahdhoo	Automatic Weather Station
Gaafu Dhaalu	Rodhavarrehaa	Automatic Weather Station
Haa Alifu	Uligamu	Automatic Weather Station
Haa Dhaalu	Hanimaadhoo	Meteorological Observation Station
Laamu	Isdhoo	Automatic Weather Station
Laamu	Kadhdhoo	Meteorological Observation Station
North Malé	Banyan Tree	Automatic Weather Station
North Malé	Malé	Meteorological Observation Station
Raa	Vadhoo	Automatic Weather Station
Thaa	Hirilandhoo	Automatic Weather Station
Vaavu	Fulidhoo	Automatic Weather Station

Source: Maldives Meteorological Service, 2017.

Map V.4: Maldives, Meteorological Observation Stations

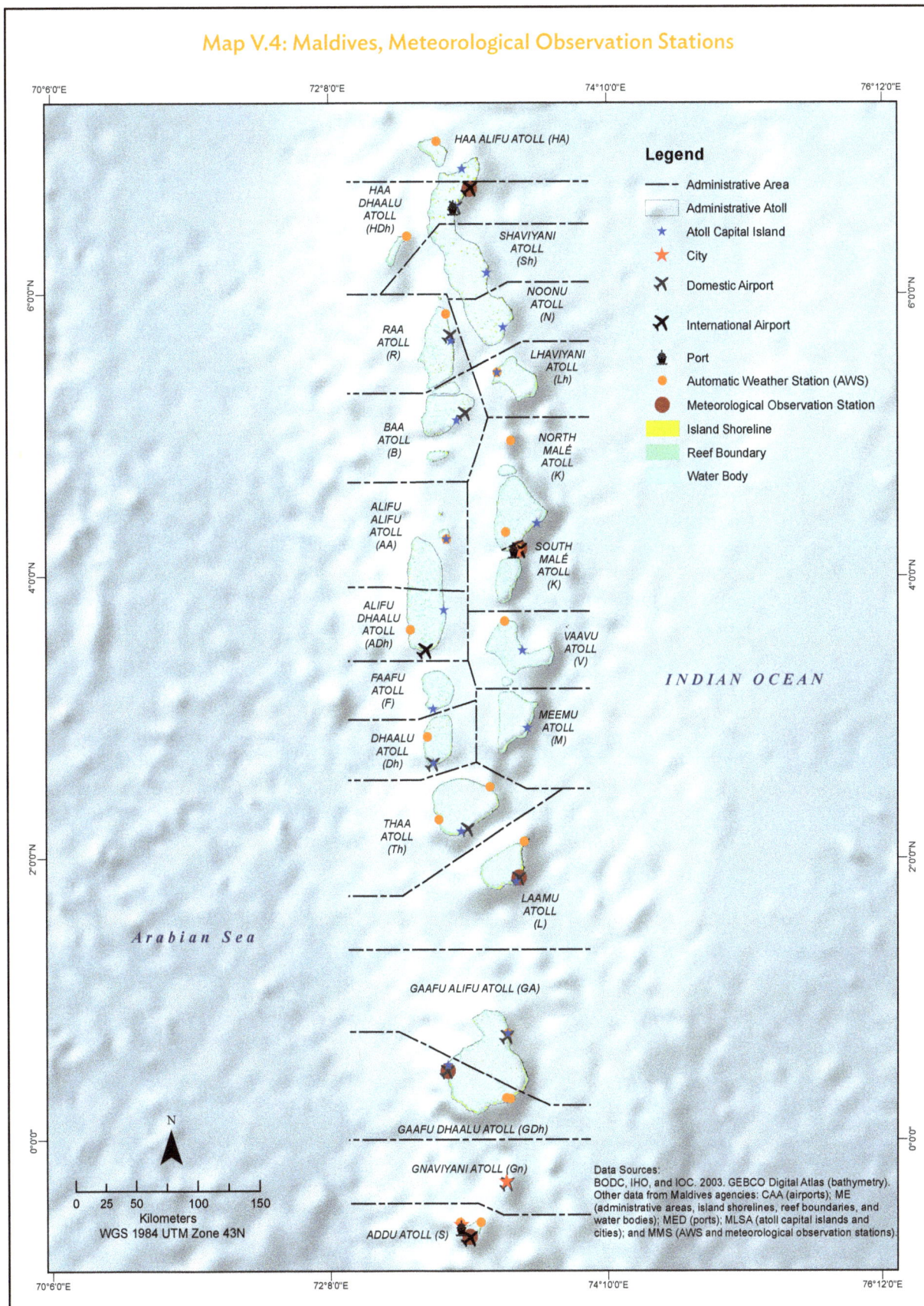

Legend

- — · — Administrative Area
- ☐ Administrative Atoll
- ★ Atoll Capital Island
- ★ City
- ✈ Domestic Airport
- ✈ International Airport
- ⚓ Port
- ● Automatic Weather Station (AWS)
- ● Meteorological Observation Station
- ▨ Island Shoreline
- ▨ Reef Boundary
- ▨ Water Body

HAA ALIFU ATOLL (HA)

HAA DHAALU ATOLL (HDh)

SHAVIYANI ATOLL (Sh)

NOONU ATOLL (N)

RAA ATOLL (R)

LHAVIYANI ATOLL (Lh)

BAA ATOLL (B)

NORTH MALÉ ATOLL (K)

ALIFU ALIFU ATOLL (AA)

SOUTH MALÉ ATOLL (K)

ALIFU DHAALU ATOLL (ADh)

VAAVU ATOLL (V)

FAAFU ATOLL (F)

MEEMU ATOLL (M)

DHAALU ATOLL (Dh)

THAA ATOLL (Th)

LAAMU ATOLL (L)

INDIAN OCEAN

Arabian Sea

GAAFU ALIFU ATOLL (GA)

GAAFU DHAALU ATOLL (GDh)

GNAVIYANI ATOLL (Gn)

ADDU ATOLL (S)

N

0 25 50   100   150
Kilometers
WGS 1984 UTM Zone 43N

Data Sources:
BODC, IHO, and IOC. 2003. GEBCO Digital Atlas (bathymetry).
Other data from Maldives agencies: CAA (airports); ME
(administrative areas, island shorelines, reef boundaries, and
water bodies); MED (ports); MLSA (atoll capital islands and
cities); and MMS (AWS and meteorological observation stations).

Historical Annual Climate

Daily rainfall and temperature data for the baseline years of 1970–2005 from the Global Historical Climatology Network (GHCN) were averaged and interpolated using thin plate splines for historical annual and seasonal climate mapping.

Rainfall

The resulting annual average rainfall map for 1970–2005 (Map V.5) shows that the least amount of rainfall is received by the atolls in the middle portion of Maldives, from Baa Atoll down to Gnaviyani. These atolls receive an average daily rainfall of 2.1–3.4 millimeters a day (mm/day) and face a greater challenge in sourcing their water supply. On the contrary, rainfall is highest in the northern atolls (Haa Alifu and Haa Dhaalu). These atolls receive an average of 4.6–5.8 mm/day. The rest of the atolls receive an average of 3.4–4.6 mm/day.

Temperature

The annual average temperature map for 1970–2005 (Map V.6) illustrates a slight difference in average annual temperature across Maldives. Higher latitudes have slightly higher average annual temperature (28.0°C–28.4°C) compared with the southern atolls. Southern atolls—from Meemu Atoll and Dhaalu Atoll down to Addu Atoll—have an average annual temperature of 27.7°C–28.0°C.

Sunset view from the Malé fishport. Fishing is the dominant economic activity and largest sector in Maldives, followed by tourism. Trends in these industries can be affected by the country's tropical

Map V.5: Maldives, Annual Average Rainfall (1970-2005)

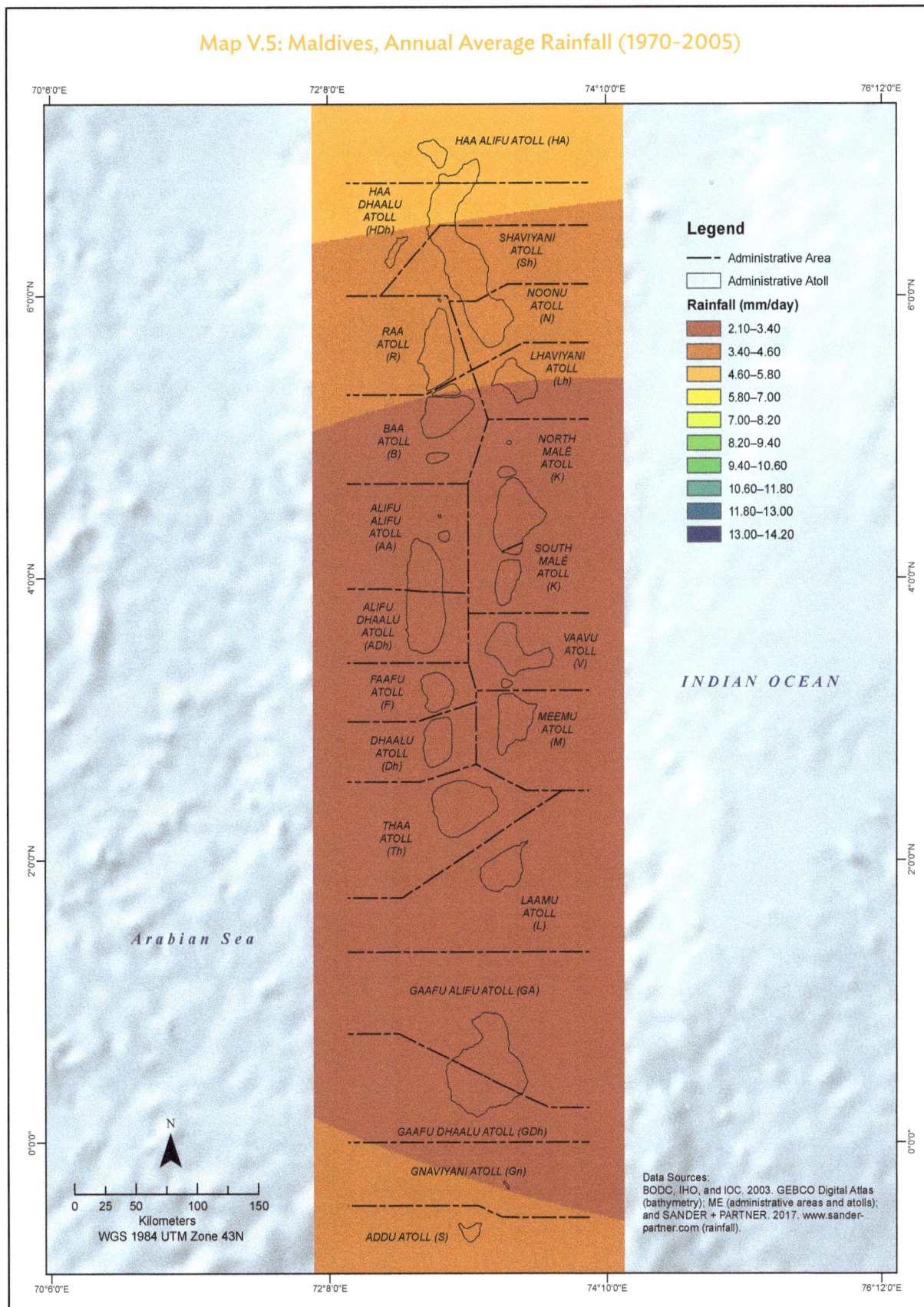

Legend

- – – Administrative Area
- ☐ Administrative Atoll

Rainfall (mm/day)

■	2.10–3.40
■	3.40–4.60
■	4.60–5.80
■	5.80–7.00
■	7.00–8.20
■	8.20–9.40
■	9.40–10.60
■	10.60–11.80
■	11.80–13.00
■	13.00–14.20

INDIAN OCEAN

Arabian Sea

N

0 25 50 100 150
Kilometers
WGS 1984 UTM Zone 43N

Atolls labeled: HAA ALIFU ATOLL (HA), HAA DHAALU ATOLL (HDh), SHAVIYANI ATOLL (Sh), NOONU ATOLL (N), RAA ATOLL (R), LHAVIYANI ATOLL (Lh), BAA ATOLL (B), NORTH MALÉ ATOLL (K), ALIFU ALIFU ATOLL (AA), SOUTH MALÉ ATOLL (K), ALIFU DHAALU ATOLL (ADh), FAAFU ATOLL (F), VAAVU ATOLL (V), MEEMU ATOLL (M), DHAALU ATOLL (Dh), THAA ATOLL (Th), LAAMU ATOLL (L), GAAFU ALIFU ATOLL (GA), GAAFU DHAALU ATOLL (GDh), GNAVIYANI ATOLL (Gn), ADDU ATOLL (S)

Data Sources:
BODC, IHO, and IOC. 2003. GEBCO Digital Atlas (bathymetry); ME (administrative areas and atolls); and SANDER + PARTNER. 2017. www.sander-partner.com (rainfall).

Map V.6: Maldives, Annual Average Temperature (1970–2005)

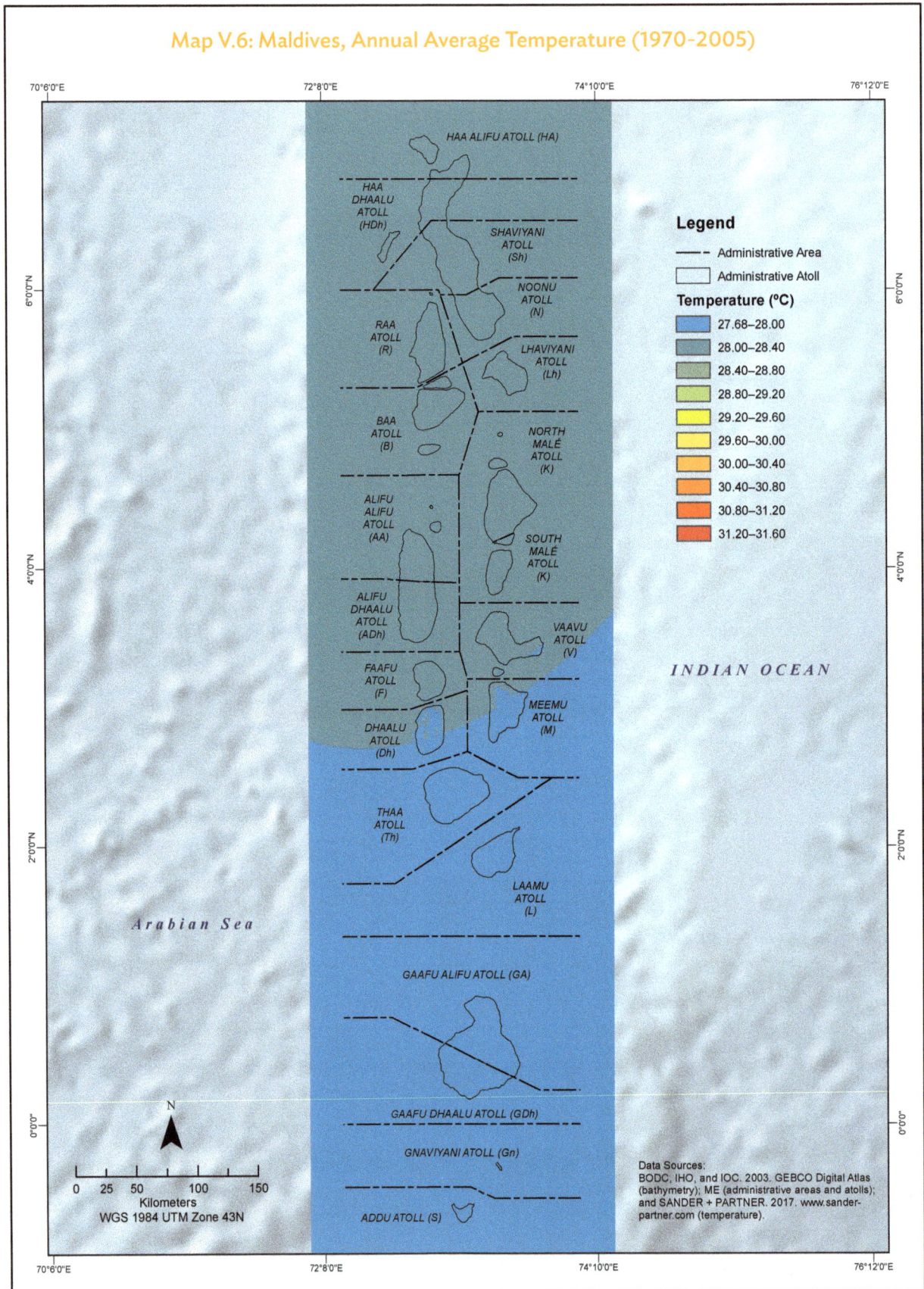

HAA ALIFU ATOLL (HA)

HAA DHAALU ATOLL (HDh)

SHAVIYANI ATOLL (Sh)

NOONU ATOLL (N)

RAA ATOLL (R)

LHAVIYANI ATOLL (Lh)

BAA ATOLL (B)

NORTH MALÉ ATOLL (K)

ALIFU ALIFU ATOLL (AA)

SOUTH MALÉ ATOLL (K)

ALIFU DHAALU ATOLL (ADh)

VAAVU ATOLL (V)

FAAFU ATOLL (F)

MEEMU ATOLL (M)

DHAALU ATOLL (Dh)

THAA ATOLL (Th)

LAAMU ATOLL (L)

Arabian Sea

GAAFU ALIFU ATOLL (GA)

GAAFU DHAALU ATOLL (GDh)

GNAVIYANI ATOLL (Gn)

ADDU ATOLL (S)

INDIAN OCEAN

Legend

– – – Administrative Area

☐ Administrative Atoll

Temperature (°C)

▨	27.68–28.00
▨	28.00–28.40
▨	28.40–28.80
▨	28.80–29.20
▨	29.20–29.60
▨	29.60–30.00
▨	30.00–30.40
▨	30.40–30.80
▨	30.80–31.20
▨	31.20–31.60

N

0   25   50   100   150
Kilometers
WGS 1984 UTM Zone 43N

Data Sources:
BODC, IHO, and IOC. 2003. GEBCO Digital Atlas (bathymetry); ME (administrative areas and atolls); and SANDER + PARTNER. 2017. www.sander-partner.com (temperature).

Historical Seasonal Climate

For seasonal climate, the same dataset from GHCN was used. These maps are presented by season: December, January, February (DJF); March, April, May (MAM); June, July, August (JJA); and September, October, November (SON). The maps show rainfall and temperature variation across time and space.

Shoreline in a resort island. Maldives has developed numerous resort islands to serve the growing tourism industry, an important catalyst in the country's overall economic growth. The industry, however, can also be vulnerable to the impacts of changing climate (photo by Roberta Gerpacio).

Seasonal Average Rainfall (1970–2005)

☑ **Wetter north**
☑ **JJA and SON are wettest months**

The rainfall distribution pattern across Maldives varies across months and latitude. Greater rainfall is experienced in the south during DJF. During MAM, Maldives is generally dry. The climate is governed by monsoons. MAM covers the transition of a drier northeast monsoon to a wetter southwest monsoon. Rainfall starts to increase in JJA, during which rainfall is greater in the northern atolls and the country experiences rainfall from the southwest monsoon. The southwest monsoon continues to bring rainfall until October. The wettest months are SON—northern and southern atolls experience greater rainfall during these months.

Map V.7: Maldives, Seasonal Average Rainfall (1970–2005)

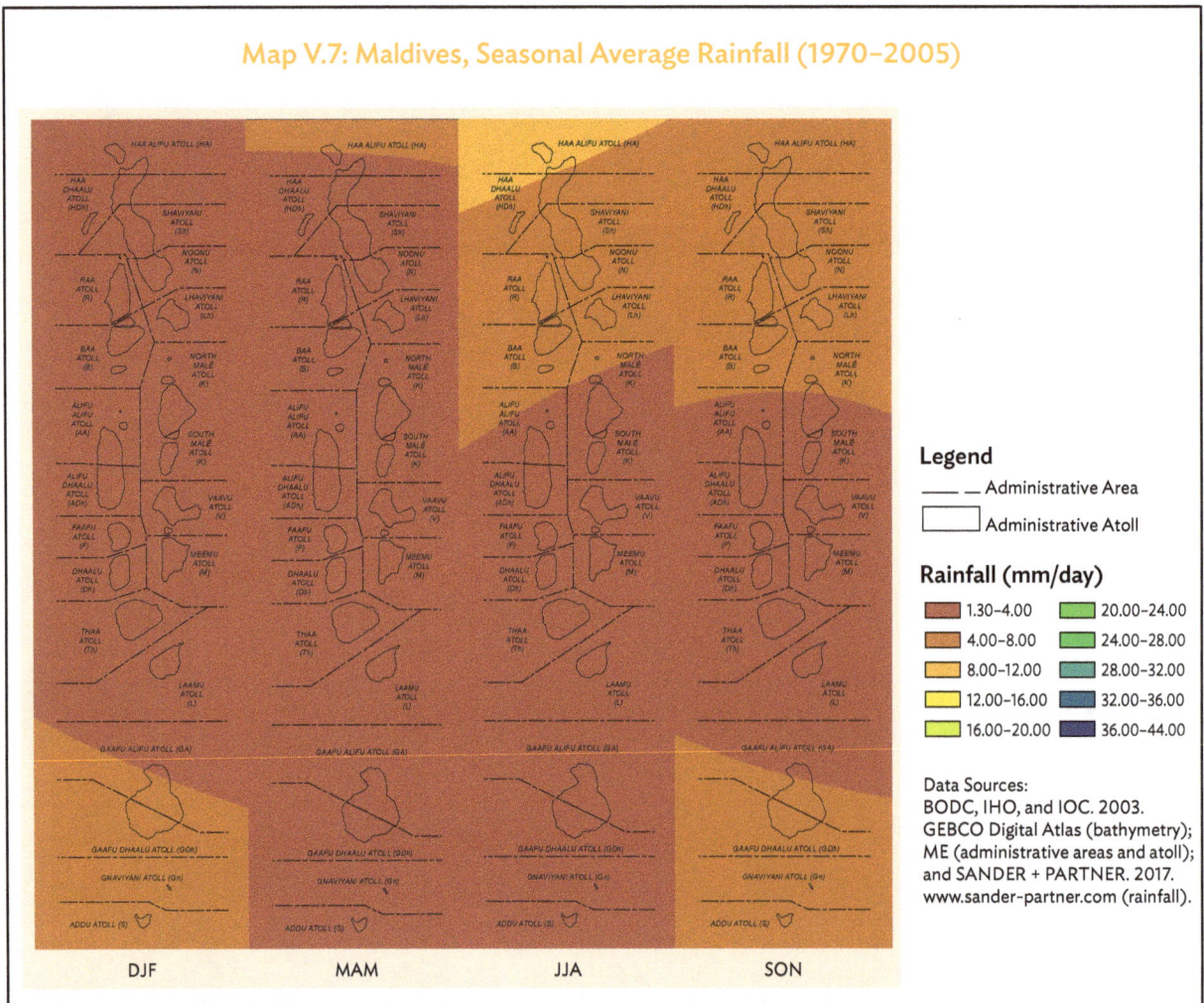

DJF   MAM   JJA   SON

Legend

——— Administrative Area

☐ Administrative Atoll

Rainfall (mm/day)

1.30–4.00	20.00–24.00
4.00–8.00	24.00–28.00
8.00–12.00	28.00–32.00
12.00–16.00	32.00–36.00
16.00–20.00	36.00–44.00

Data Sources:
BODC, IHO, and IOC. 2003.
GEBCO Digital Atlas (bathymetry);
ME (administrative areas and atoll);
and SANDER + PARTNER. 2017.
www.sander-partner.com (rainfall).

Seasonal Average Temperature (1970–2005)

Maldives is coldest during DJF and SON. The average temperature, ranging 27.40°C–27.95°C, is consistent across the country. The temperature (27.95°C–30.15°C) during MAM increases. MAM are the warmest months especially in the northern atolls. During JJA, the temperature decreases to 27.40°C–27.95 °C in the southern atolls and 27.95°C–28.50 °C in the northern and central atolls.

☑ **Warmer north**
☑ **MAM are the warmest months**

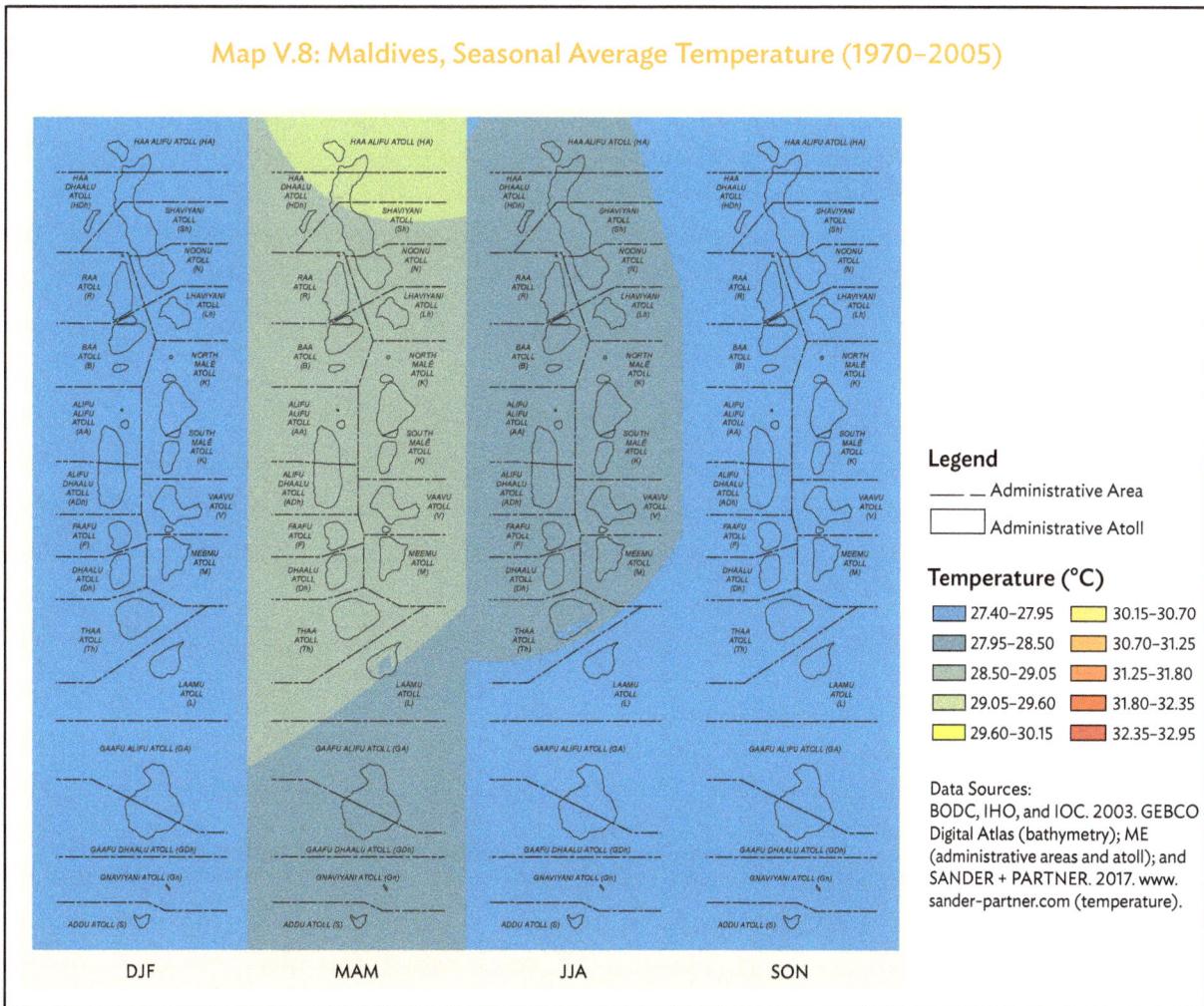

Map V.8: Maldives, Seasonal Average Temperature (1970–2005)

Legend

— Administrative Area

☐ Administrative Atoll

Temperature (°C)

27.40–27.95	30.15–30.70
27.95–28.50	30.70–31.25
28.50–29.05	31.25–31.80
29.05–29.60	31.80–32.35
29.60–30.15	32.35–32.95

Data Sources:
BODC, IHO, and IOC. 2003. GEBCO Digital Atlas (bathymetry); ME (administrative areas and atoll); and SANDER + PARTNER. 2017. www.sander-partner.com (temperature).

DJF   MAM   JJA   SON

Future Climate

Throughout the formation of our planet, the climate has changed and will continue to change. Together with geological processes, climate change continues to shape Maldives. While the historical climate of the country sustained life, future atmospheric conditions may pose several threats to life in these tropical islands.

The same data set from GHCN was downscaled to the regional climate models to have a finer resolution of 50 kilometers. The model followed the representative concentration pathways (RCPs) based on the Intergovernmental Panel on Climate Change's Fifth Assessment Report (IPCC 2014). Two RCPs (4.5 and 8.5) were selected. RCP 4.5 represents a moderate greenhouse gas emission scenario, while RCP 8.5 represents a high greenhouse gas emission scenario with no mitigation efforts. Results were further downscaled to 900 meters.

The downscaled projected climate shows a warmer and wetter Maldives in the future. The succeeding maps will illustrate the projected rainfall and temperature per decade to give a clearer picture of the changing climate.

Lone tree. A leafless tree standing in a rocky shore in Maldives. Future changes in climate will have a tremendous impact on the life cycle of trees and other forms of living organisms (photo by Ahmed Shareef).

Annual Average Rainfall (RCP 4.5)

23

According to the projected rainfall based on RCP 4.5 (moderate greenhouse gas emissions), Haa Alifu and Shaviyani atolls as well as Baa and Lhaviyani atolls will experience an increase in average annual rainfall rate in the coming decades. The maps show a slight decrease in the area with the 1.3-4 mm/day rainfall rate, indicating a general increase in rainfall rate through the decades. A southward trend in increased rainfall can also be observed in the projected average annual rainfall across the decades.

☑ **Wetter north**
☑ **Northern atolls will be wetter in 2030s and 2040s.**

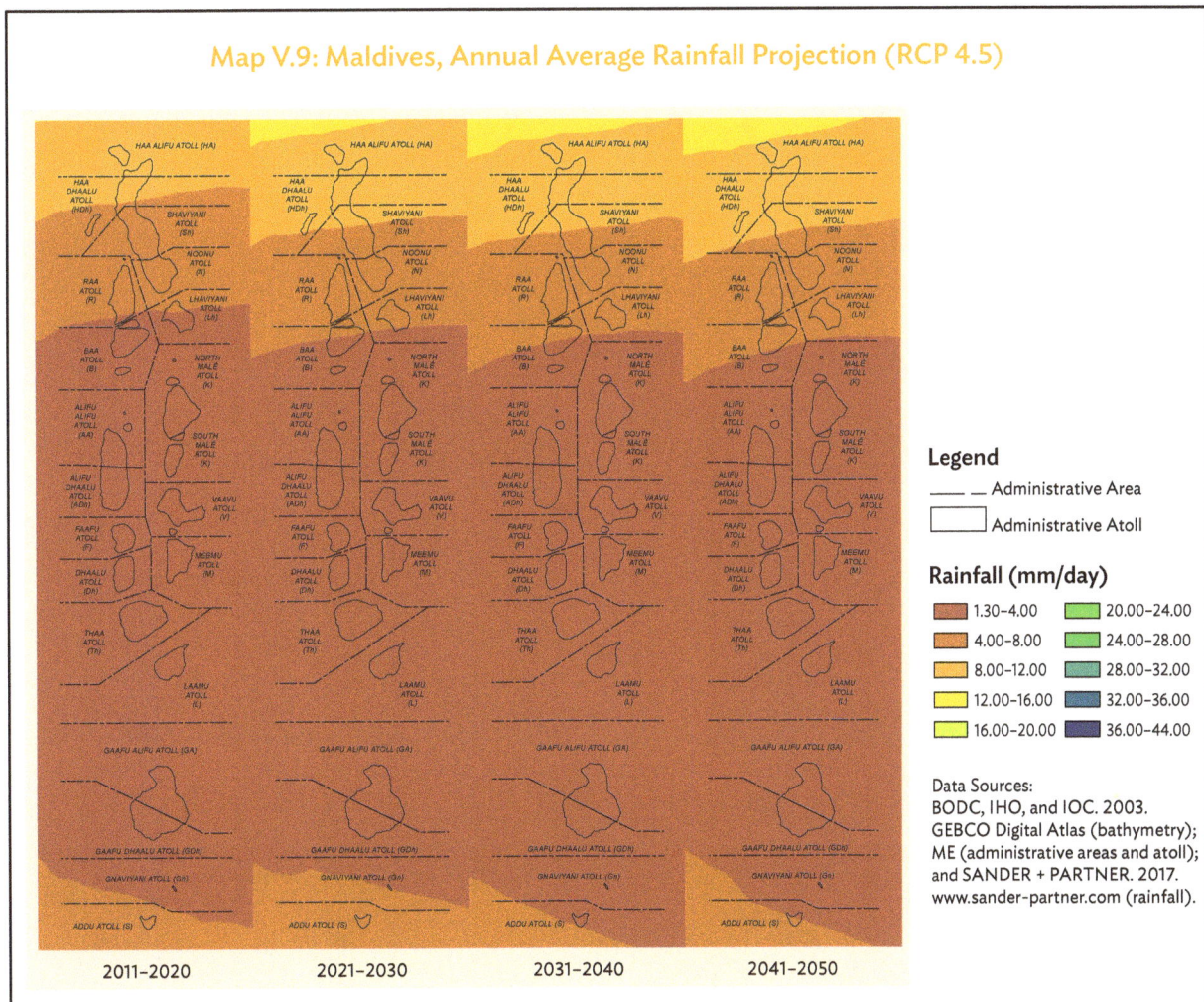

Map V.9: Maldives, Annual Average Rainfall Projection (RCP 4.5)

Legend

— — Administrative Area

☐ Administrative Atoll

Rainfall (mm/day)

1.30–4.00	20.00–24.00
4.00–8.00	24.00–28.00
8.00–12.00	28.00–32.00
12.00–16.00	32.00–36.00
16.00–20.00	36.00–44.00

Data Sources:
BODC, IHO, and IOC. 2003.
GEBCO Digital Atlas (bathymetry);
ME (administrative areas and atoll);
and SANDER + PARTNER. 2017.
www.sander-partner.com (rainfall).

2011–2020 2021–2030 2031–2040 2041–2050

Average Seasonal Rainfall Projection (DJF, RCP 4.5)

☑ **Wetter south**

☑ **Southern atolls will be wetter beginning in the 2030s**

The northeast monsoon prevails during DJF, keeping rainfall in the northern and middle portion of Maldives low (1.3–4 mm/day). These months bring rainfall (4–8 mm/day) to the southern atolls. The southern atolls will experience wetter 2030s and 2040s.

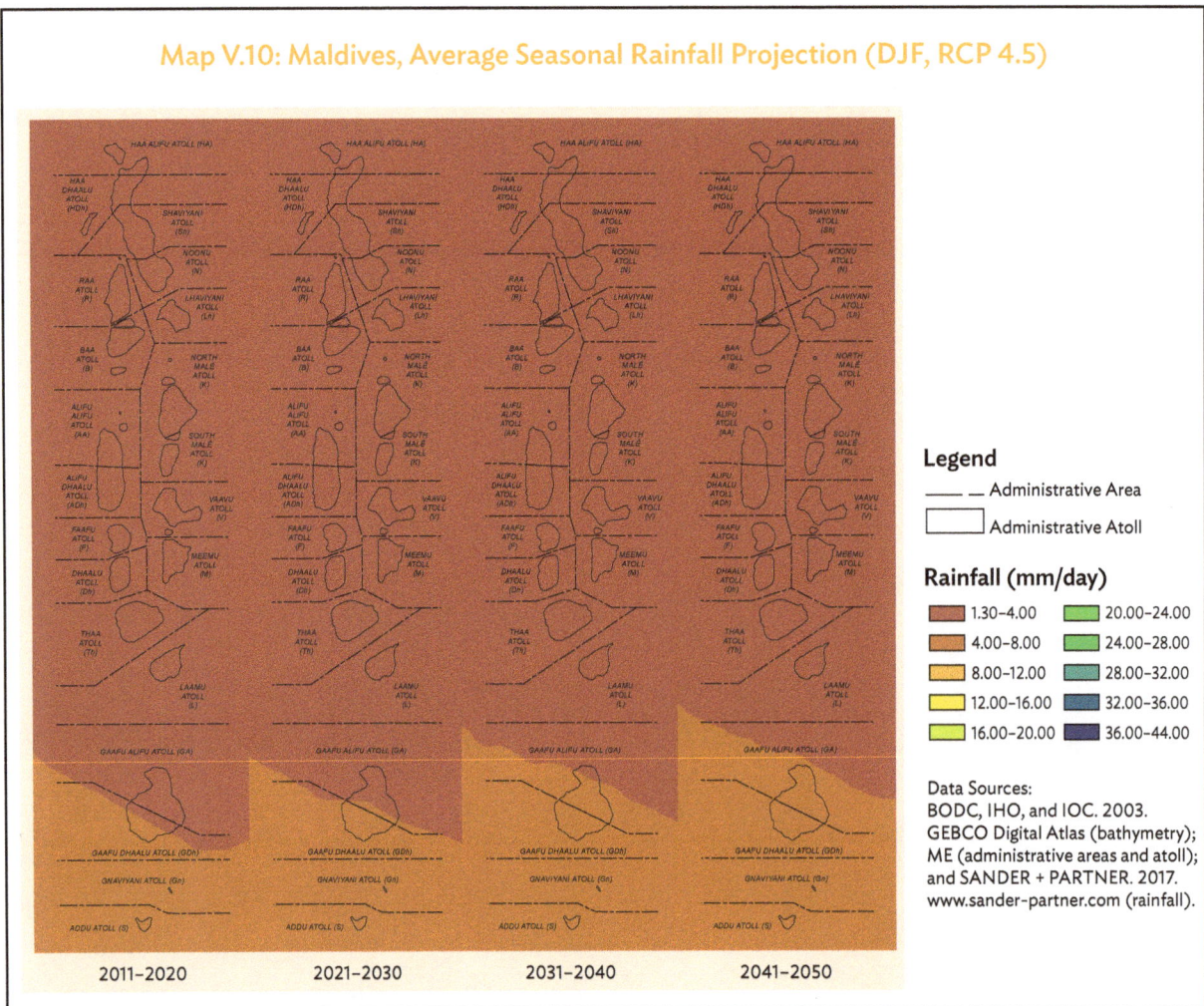

Map V.10: Maldives, Average Seasonal Rainfall Projection (DJF, RCP 4.5)

2011–2020   2021–2030   2031–2040   2041–2050

Legend

— — Administrative Area

☐ Administrative Atoll

Rainfall (mm/day)

1.30–4.00	20.00–24.00
4.00–8.00	24.00–28.00
8.00–12.00	28.00–32.00
12.00–16.00	32.00–36.00
16.00–20.00	36.00–44.00

Data Sources:
BODC, IHO, and IOC. 2003.
GEBCO Digital Atlas (bathymetry);
ME (administrative areas and atoll);
and SANDER + PARTNER. 2017.
www.sander-partner.com (rainfall).

Average Seasonal Rainfall Projection (MAM, RCP 4.5)

Maldives experiences the driest days during MAM, as the northeast monsoon is strongest these months. Rainfall, at a rate of 1.3–4 mm/day, is distributed evenly throughout the atolls during MAM and throughout the projected time periods (with the exception of Haa Alifu Atoll in the 2030s).

☑ **Generally dry**
☑ **Driest season**
☑ **No significant change across the decades**

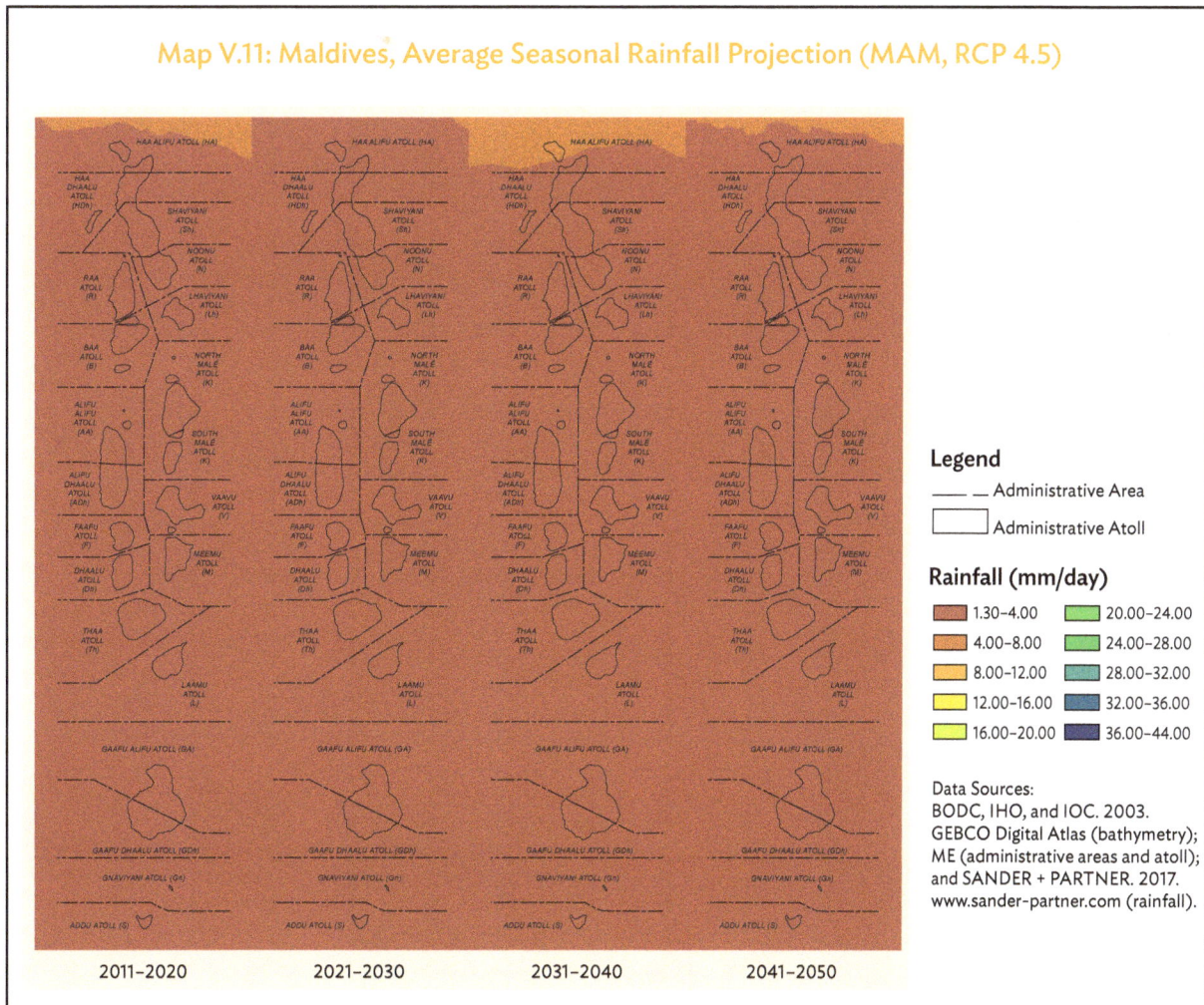

Map V.11: Maldives, Average Seasonal Rainfall Projection (MAM, RCP 4.5)

Legend

— Administrative Area
☐ Administrative Atoll

Rainfall (mm/day)

1.30–4.00	20.00–24.00
4.00–8.00	24.00–28.00
8.00–12.00	28.00–32.00
12.00–16.00	32.00–36.00
16.00–20.00	36.00–44.00

Data Sources:
BODC, IHO, and IOC. 2003.
GEBCO Digital Atlas (bathymetry);
ME (administrative areas and atoll);
and SANDER + PARTNER. 2017.
www.sander-partner.com (rainfall).

2011–2020   2021–2030   2031–2040   2041–2050

Average Seasonal Rainfall Projection (JJA, RCP 4.5)

☑ **Wetter north**

☑ **Wet season**

☑ **Wetter middle atolls in the 2030s and 2040s**

Maldives experiences the highest rainfall rate during JJA compared with other seasons. In these months, rainfall is observed to increase, especially in the northern atolls as the southwest monsoon intensifies. Rainfall rate greater than 4 mm/day is expected to move southward in the 2030s, reaching the atolls in the middle portion of Maldives.

Map V.12: Maldives, Average Seasonal Rainfall Projection (JJA, RCP 4.5)

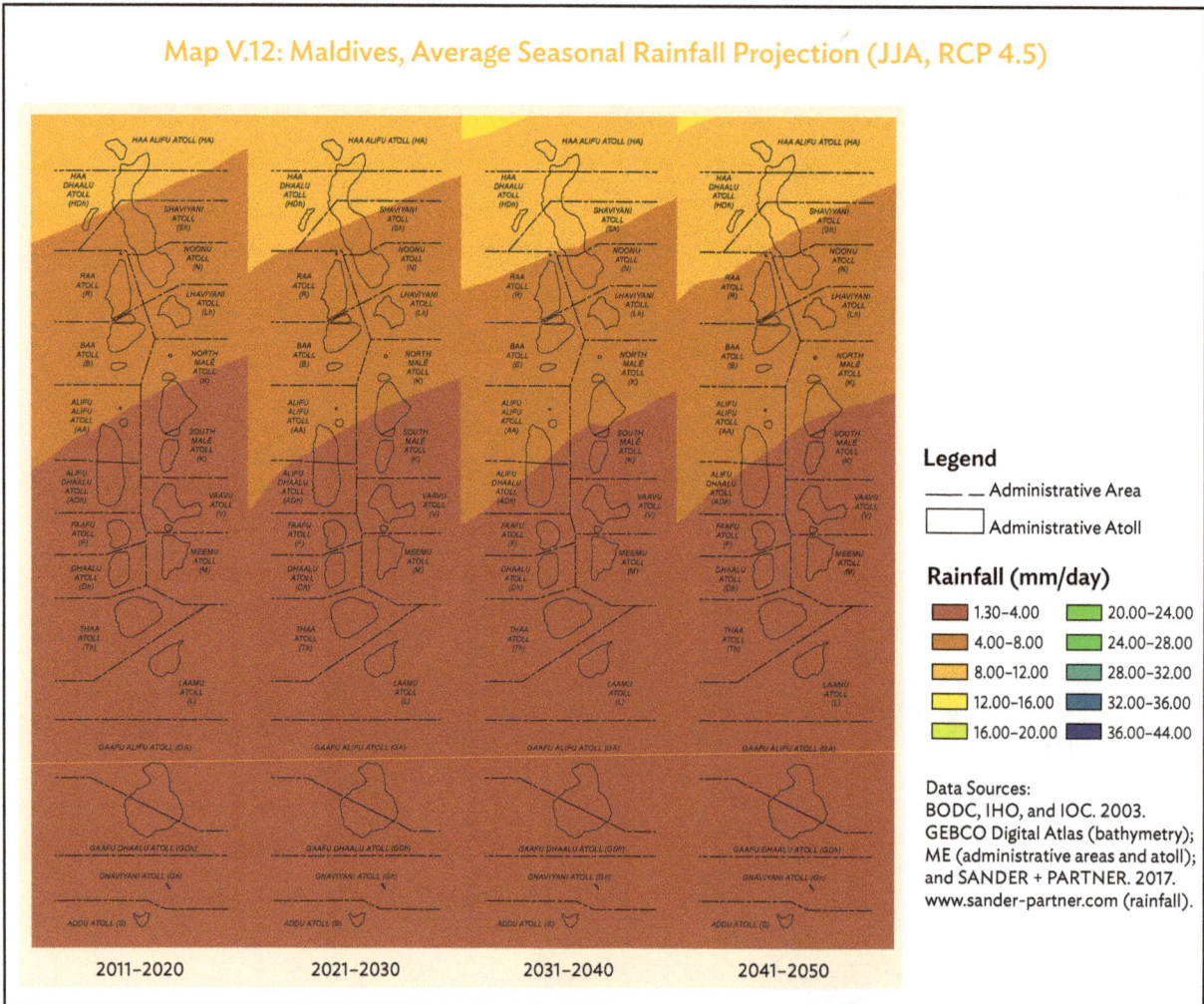

Legend

— — Administrative Area

☐ Administrative Atoll

Rainfall (mm/day)

1.30–4.00	20.00–24.00
4.00–8.00	24.00–28.00
8.00–12.00	28.00–32.00
12.00–16.00	32.00–36.00
16.00–20.00	36.00–44.00

Data Sources:
BODC, IHO, and IOC. 2003.
GEBCO Digital Atlas (bathymetry);
ME (administrative areas and atoll);
and SANDER + PARTNER. 2017.
www.sander-partner.com (rainfall).

2011–2020　2021–2030　2031–2040　2041–2050

Average Seasonal Rainfall Projection (SON, RCP 4.5)

During SON, the southwest monsoon starts to weaken, bringing the rainfall rate down to less than 8 mm/day. The maps presenting the time series of seasonal rainfall for SON consistently show that across the projected time periods, the southern half of Maldives receives minimal (1.3–4 mm/day) rainfall while the northern half of Maldives experiences a higher rainfall rate (4–8 mm/day).

- ☑ **Wetter north**
- ☑ **Wet season**
- ☑ **No significant shift in rainfall distribution across decades**

Map V.13: Maldives, Average Seasonal Rainfall Projection (SON, RCP 4.5)

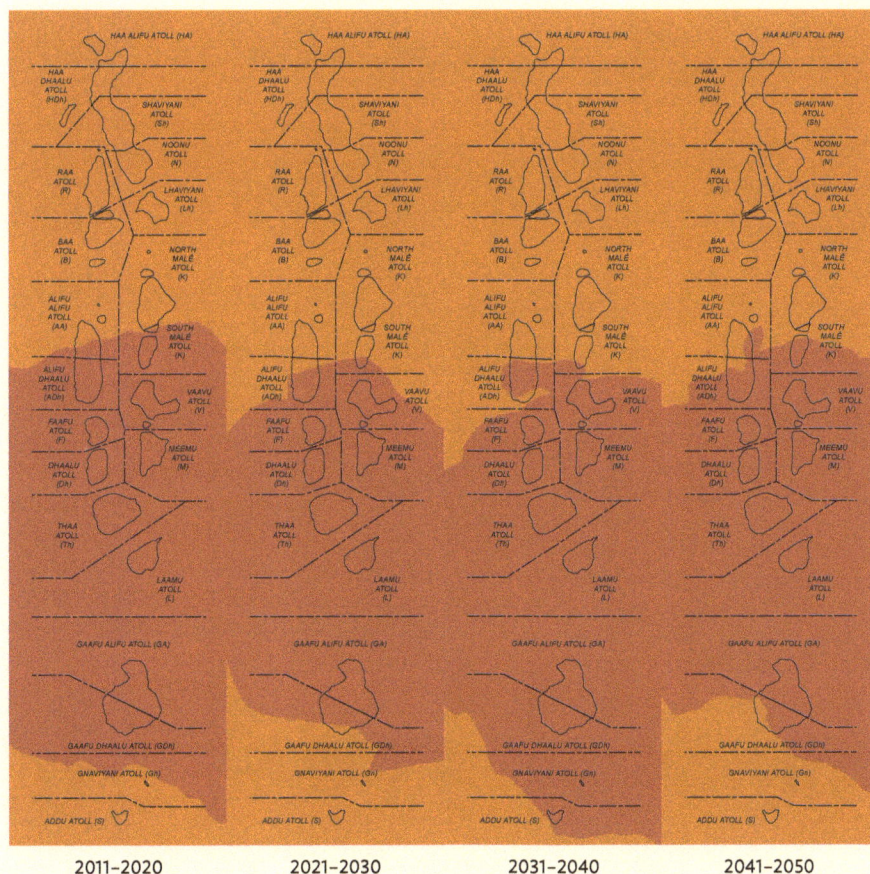

Legend

— — Administrative Area

☐ Administrative Atoll

Rainfall (mm/day)

1.30–4.00	20.00–24.00
4.00–8.00	24.00–28.00
8.00–12.00	28.00–32.00
12.00–16.00	32.00–36.00
16.00–20.00	36.00–44.00

Data Sources:
BODC, IHO, and IOC. 2003.
GEBCO Digital Atlas (bathymetry);
ME (administrative areas and atoll);
and SANDER + PARTNER. 2017.
www.sander-partner.com (rainfall).

2011–2020   2021–2030   2031–2040   2041–2050

Average Annual Rainfall Projection (RCP 8.5)

☑ **Wetter north**

☑ **North will be wetter in the future**

☑ **No significant difference with RCP 4.5**

RCP 8.5 depicts a scenario in which greenhouse gas emissions are higher. As a result, the change in rainfall would also be greater compared with RCP 4.5. However, when comparing the RCP 8.5 and RCP 4.5 maps, there is minimal visible difference in the average rainfall. In this time series of annual rainfall projection for RCP 8.5, an increasing rainfall rate trend from 2011 to 2050 is visible. Higher rainfall rate (4–16 mm/day) will be experienced by atolls in northern Maldives. However, as the years progress, more atolls in the north will experience wetter days.

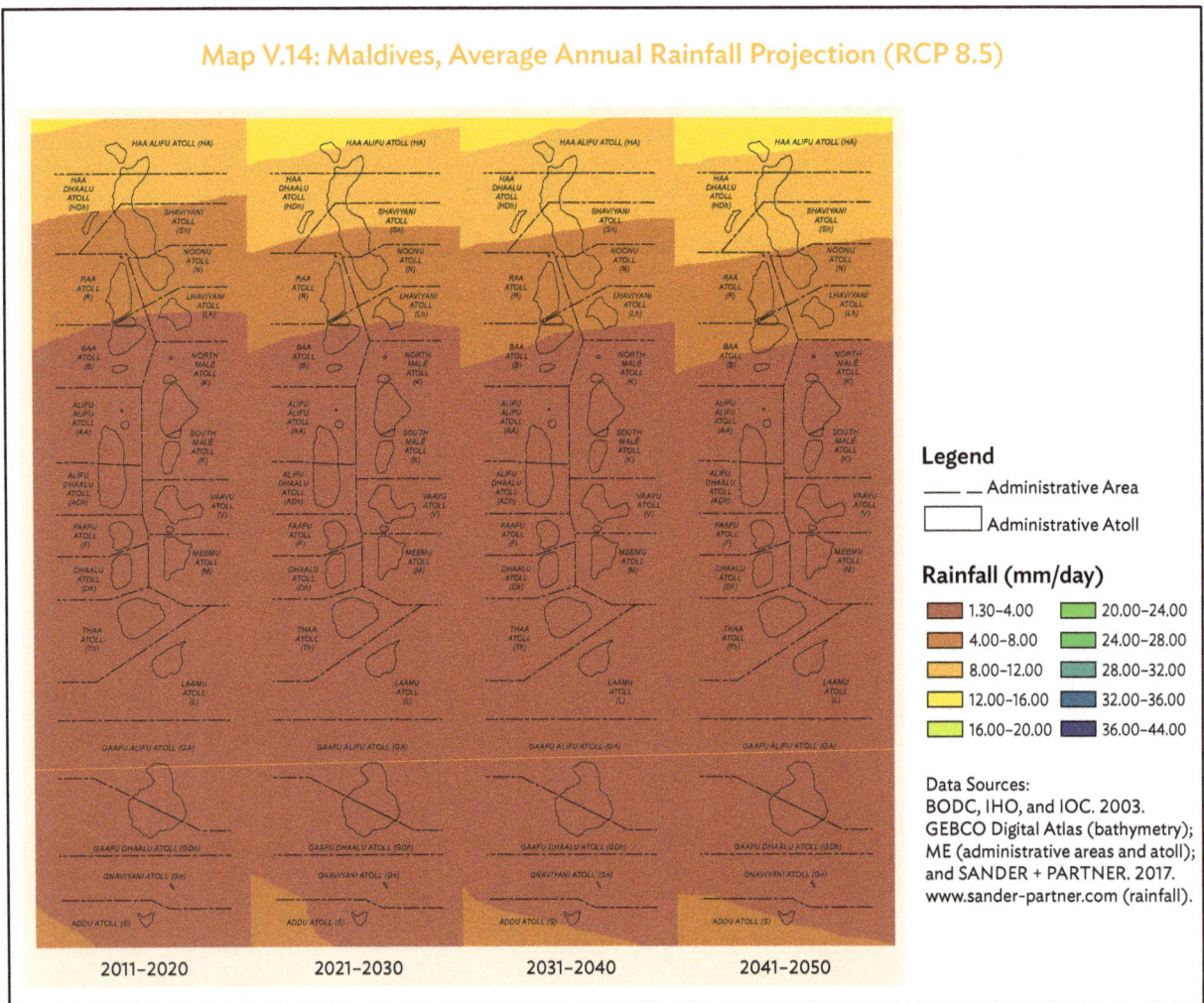

Map V.14: Maldives, Average Annual Rainfall Projection (RCP 8.5)

Legend

— — Administrative Area

☐ Administrative Atoll

Rainfall (mm/day)

1.30–4.00	20.00–24.00
4.00–8.00	24.00–28.00
8.00–12.00	28.00–32.00
12.00–16.00	32.00–36.00
16.00–20.00	36.00–44.00

Data Sources:
BODC, IHO, and IOC. 2003.
GEBCO Digital Atlas (bathymetry);
ME (administrative areas and atoll);
and SANDER + PARTNER. 2017.
www.sander-partner.com (rainfall).

2011–2020   2021–2030   2031–2040   2041–2050

Average Seasonal Rainfall Projection (DJF, RCP 8.5)

RCP 8.5 and RCP 4.5 projections of average seasonal rainfall show the same picture with a slight difference. The south of Maldives will remain wetter compared to the north. More atolls in the south will experience a higher rainfall rate (4–8 mm/day). However, there is a slight increase in the area covered by the 4–8 mm/day rainfall rate for RCP 8.5 as compared to the RCP 4.5 projection.

- ☑ Wetter south
- ☑ More atolls in the south will experience higher rainfall rate
- ☑ 4–8 mm/day rainfall rate has greater coverage than RCP 4.5 projection

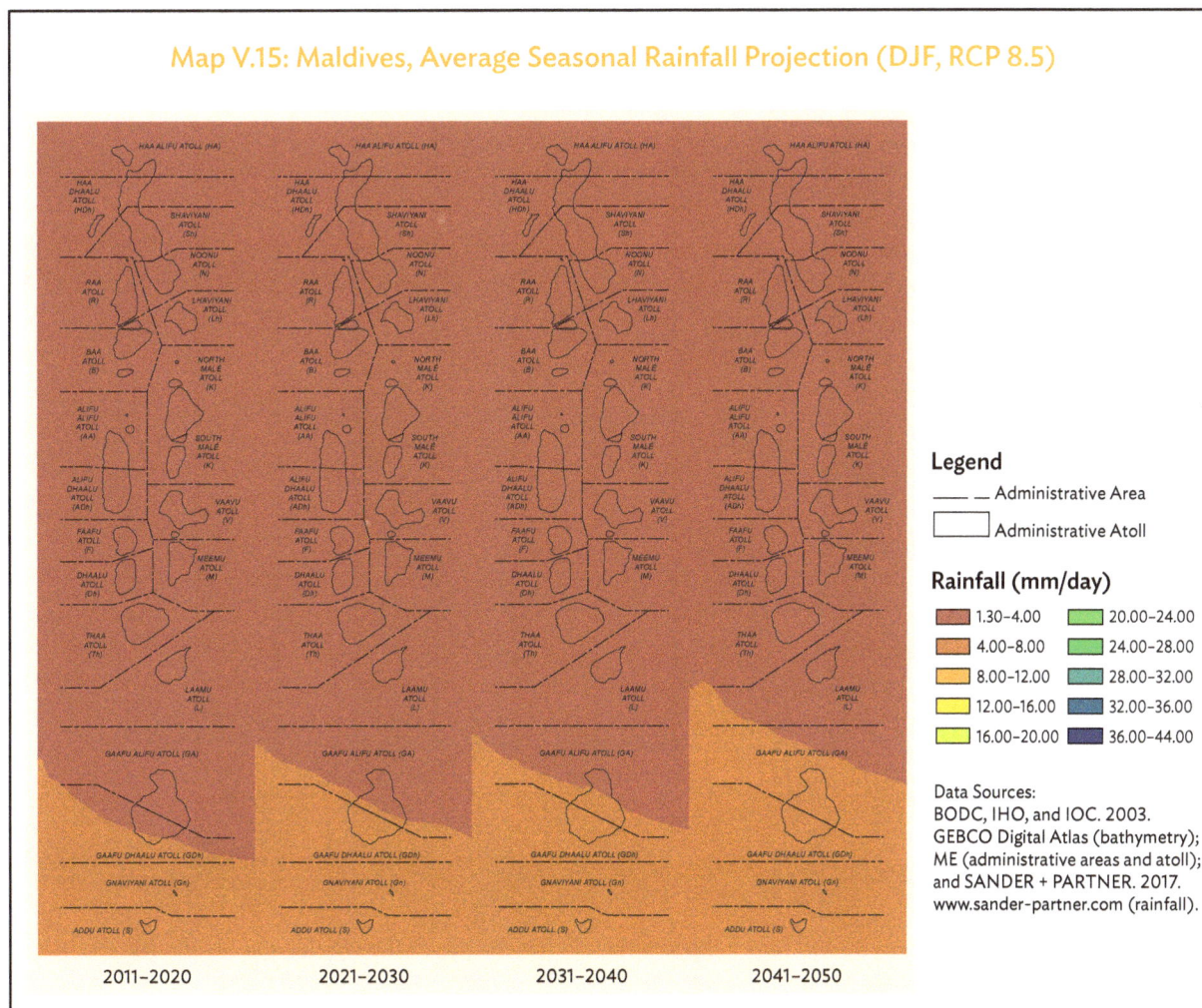

Map V.15: Maldives, Average Seasonal Rainfall Projection (DJF, RCP 8.5)

Legend

— — Administrative Area

☐ Administrative Atoll

Rainfall (mm/day)

1.30–4.00	20.00–24.00
4.00–8.00	24.00–28.00
8.00–12.00	28.00–32.00
12.00–16.00	32.00–36.00
16.00–20.00	36.00–44.00

Data Sources:
BODC, IHO, and IOC. 2003. GEBCO Digital Atlas (bathymetry); ME (administrative areas and atoll); and SANDER + PARTNER. 2017. www.sander-partner.com (rainfall).

| 2011–2020 | 2021–2030 | 2031–2040 | 2041–2050 |

Average Seasonal Rainfall Projection (MAM, RCP 8.5)

- ☑ **Low rainfall rate is evenly distributed across Maldives**
- ☑ **No visible change in rainfall across time periods**
- ☑ **No visible difference between RCP 8.5 and RCP 4.5 projections**

The MAM average seasonal rainfall projection for RCP 8.5 is similar to that of RCP 4.5. The rainfall rate for the whole Maldivian archipelago is still at its lowest range during MAM. The low rainfall rate is sustained throughout all four time periods.

Map V.16: Maldives, Average Seasonal Rainfall Projection (MAM, RCP 8.5)

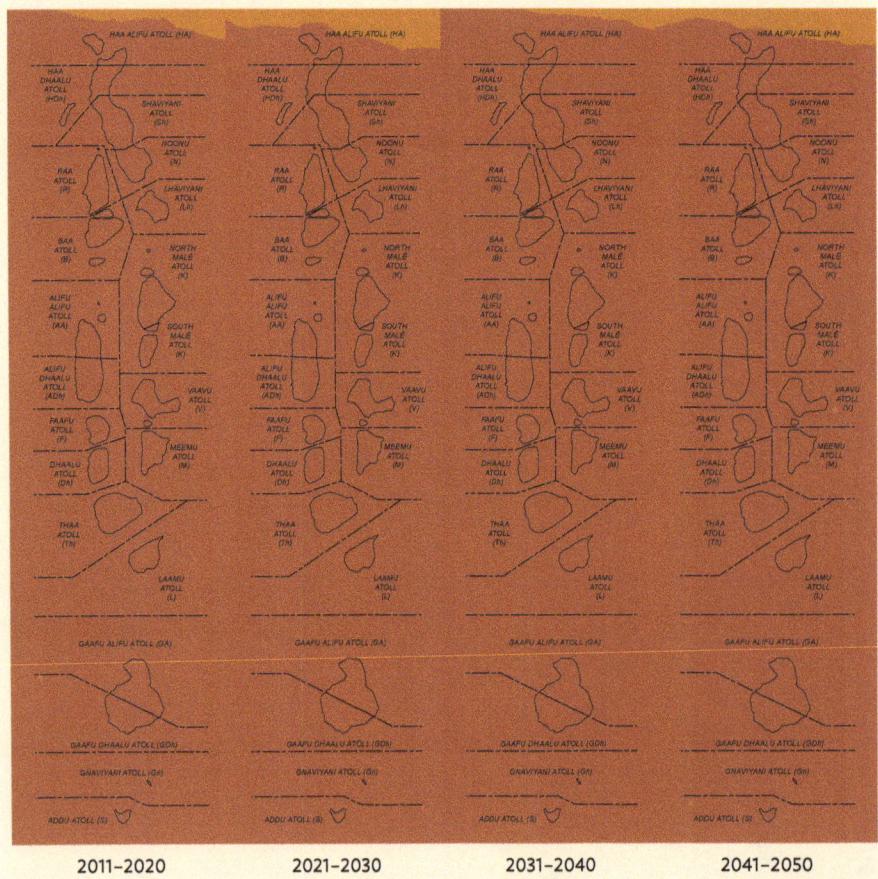

Legend

— — Administrative Area

▭ Administrative Atoll

Rainfall (mm/day)

1.30–4.00	20.00–24.00
4.00–8.00	24.00–28.00
8.00–12.00	28.00–32.00
12.00–16.00	32.00–36.00
16.00–20.00	36.00–44.00

Data Sources:
BODC, IHO, and IOC. 2003.
GEBCO Digital Atlas (bathymetry);
ME (administrative areas and atoll);
and SANDER + PARTNER. 2017.
www.sander-partner.com (rainfall).

2011–2020   2021–2030   2031–2040   2041–2050

Average Seasonal Rainfall Projection (JJA, RCP 8.5)

JJA will continue to be the wettest months of the year, especially in the northern atolls of Maldives. Haa Alifu Atoll and Haa Dhaalu Atoll will experience the highest average daily rainfall rate (up to 12 mm/day). From the 2030s to the 2040s, Shaviyani Atoll will also experience this average daily rainfall rate.

- ☑ **Wetter north**
- ☑ **Wettest season**
- ☑ **Wetter north in the future**
- ☑ **No visible difference between RCP 8.5 and RCP 4.5 projections**

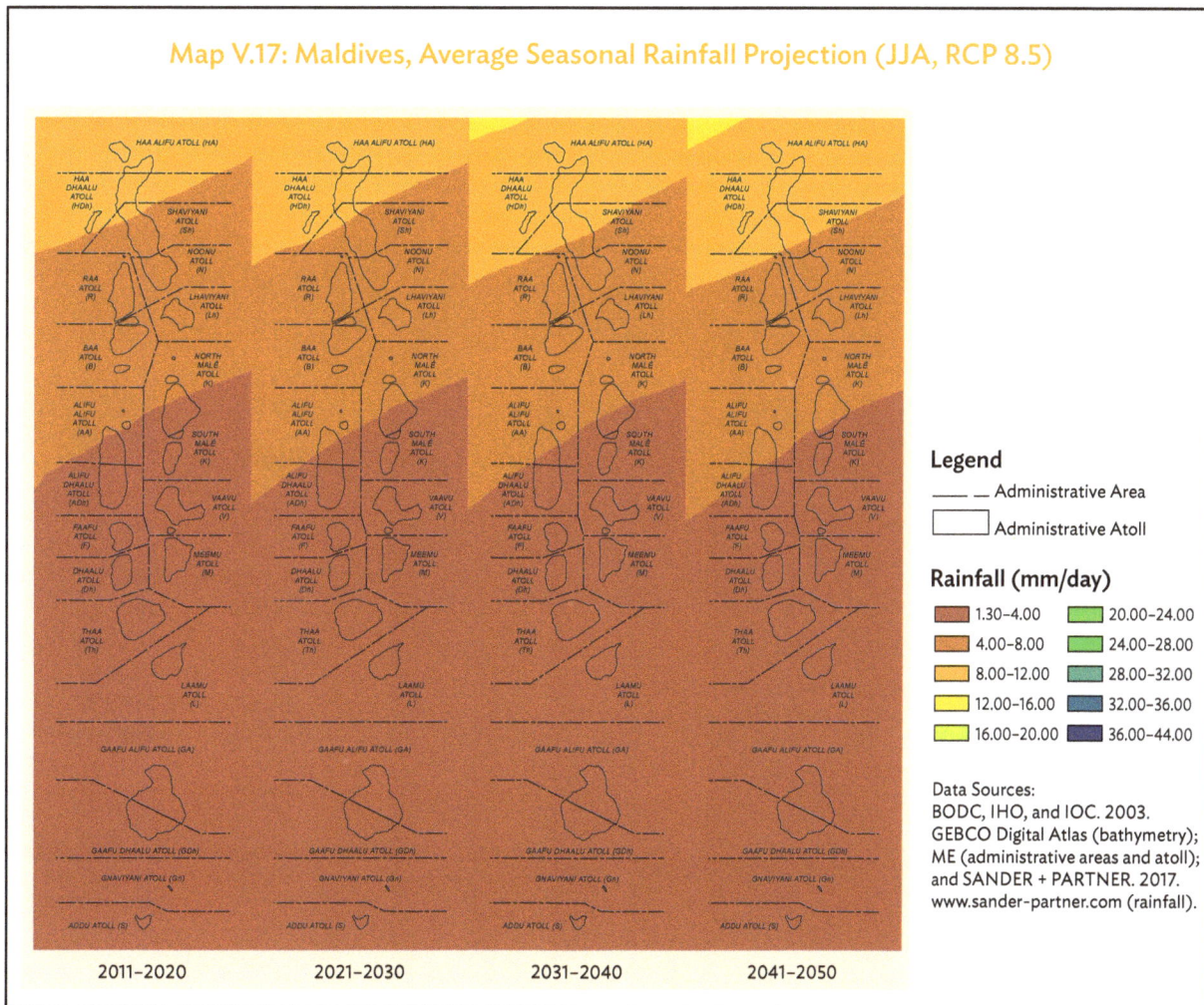

Map V.17: Maldives, Average Seasonal Rainfall Projection (JJA, RCP 8.5)

2011–2020   2021–2030   2031–2040   2041–2050

Legend

—   — Administrative Area

☐   Administrative Atoll

Rainfall (mm/day)

1.30–4.00	20.00–24.00
4.00–8.00	24.00–28.00
8.00–12.00	28.00–32.00
12.00–16.00	32.00–36.00
16.00–20.00	36.00–44.00

Data Sources:
BODC, IHO, and IOC. 2003. GEBCO Digital Atlas (bathymetry); ME (administrative areas and atoll); and SANDER + PARTNER. 2017. www.sander-partner.com (rainfall).

Average Seasonal Rainfall Projection (SON, RCP 8.5)

32

- ☑ **Wetter north**
- ☑ **More northern atolls with higher rainfall rate (4–8 mm/day) in the future**
- ☑ **Greater area with 4–8 mm/day rainfall rate in the future in RCP 8.5 compared with RCP 4.5**

The time series of projected average rainfall during SON based on RCP 8.5 shows an increasing area covered by a rainfall rate of 4–8 mm/day. This translates to increasing rainfall in the middle portion of Maldives in the 2020s. During SON, the northern half of Maldives has a rainfall rate of 4–8 mm/day and the southern half has a low rainfall rate of 1.3–4 mm/day.

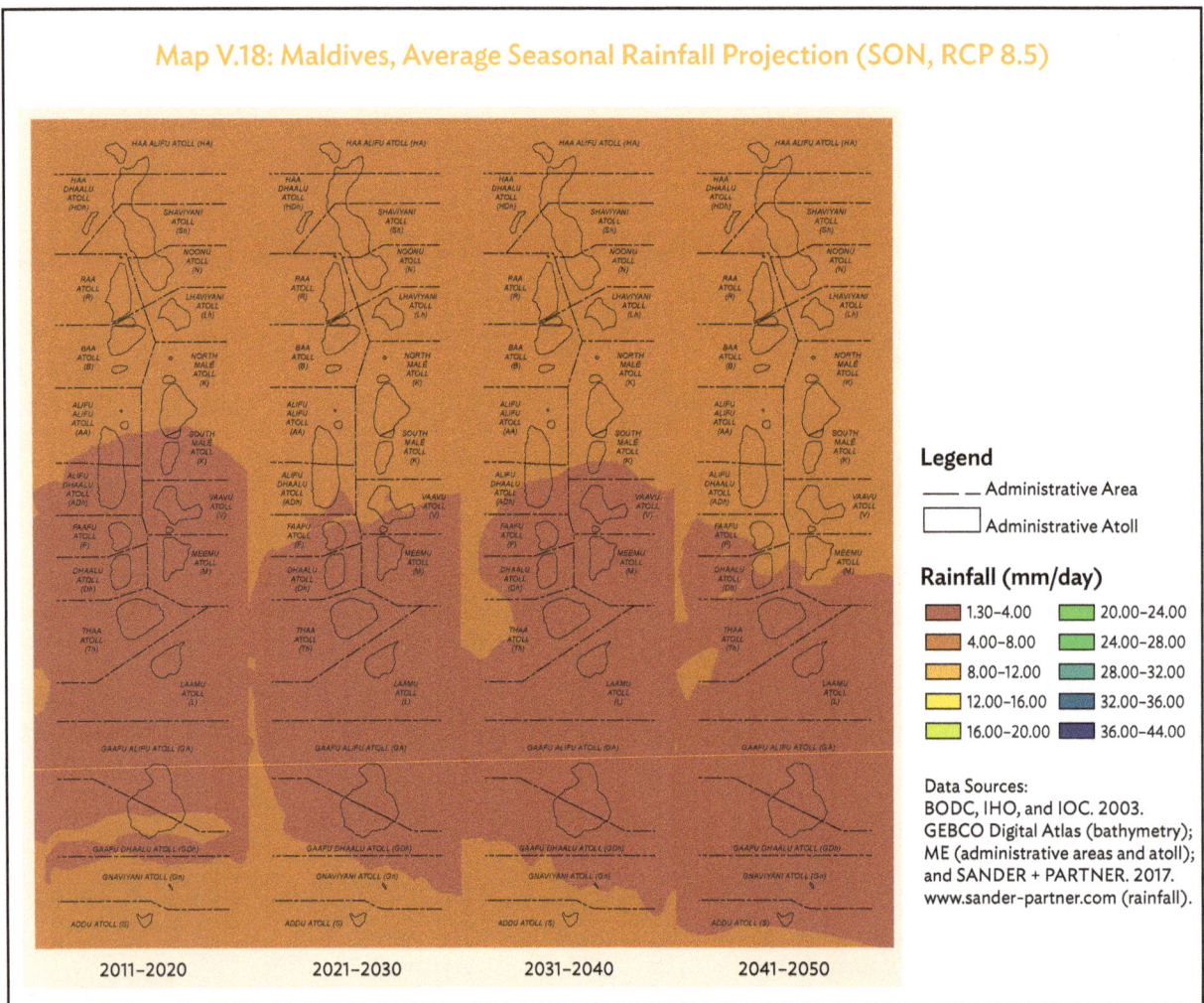

Map V.18: Maldives, Average Seasonal Rainfall Projection (SON, RCP 8.5)

Legend

— Administrative Area

☐ Administrative Atoll

Rainfall (mm/day)

1.30–4.00	20.00–24.00
4.00–8.00	24.00–28.00
8.00–12.00	28.00–32.00
12.00–16.00	32.00–36.00
16.00–20.00	36.00–44.00

Data Sources:
BODC, IHO, and IOC. 2003.
GEBCO Digital Atlas (bathymetry);
ME (administrative areas and atoll);
and SANDER + PARTNER. 2017.
www.sander-partner.com (rainfall).

Average Annual Temperature Projection (RCP 4.5)

In this decade, the average annual temperature in Maldives (with the exception of Haafu Alifu) is 28.4°C–28.8°C. Maldives will have warmer days (average of 29.05°C–30.15°C) in the 2030s and 2040s, particularly in northern Maldives where the average annual temperature will range from 29.60°C to 30.15°C.

☑ **Warmer north**
☑ **Warmer Maldives in the future**

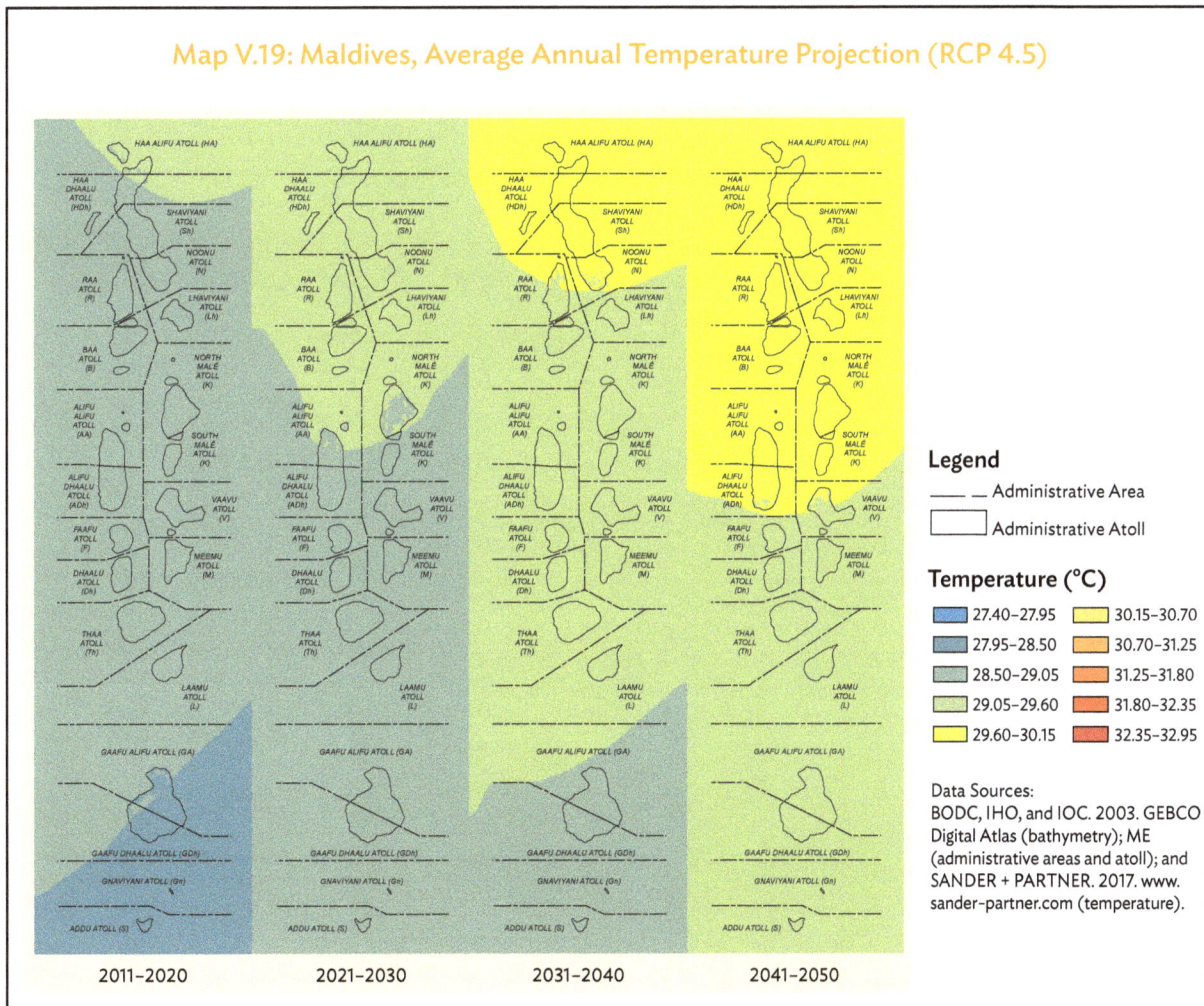

Map V.19: Maldives, Average Annual Temperature Projection (RCP 4.5)

2011–2020   2021–2030   2031–2040   2041–2050

Legend

— Administrative Area

☐ Administrative Atoll

Temperature (°C)

27.40–27.95	30.15–30.70
27.95–28.50	30.70–31.25
28.50–29.05	31.25–31.80
29.05–29.60	31.80–32.35
29.60–30.15	32.35–32.95

Data Sources:
BODC, IHO, and IOC. 2003. GEBCO Digital Atlas (bathymetry); ME (administrative areas and atoll); and SANDER + PARTNER. 2017. www.sander-partner.com (temperature).

Average Seasonal Temperature Projection (DJF, RCP 4.5)

☑ **Warmer north**
☑ **Warmer Maldives in the future**

The current average temperature during DJF in Maldives ranges from 27.95°C to 28.50°C, with higher temperature in the northeastern region. In the 2030s and 2040s, Maldives will have warmer days during DJF. People living in the north will experience an average temperature ranging from 29.05°C to 29.60°C beginning 2031. By 2050, this average temperature during DJF will be experienced across the country.

Map V.20: Maldives, Average Seasonal Temperature Projection (DJF, RCP 4.5)

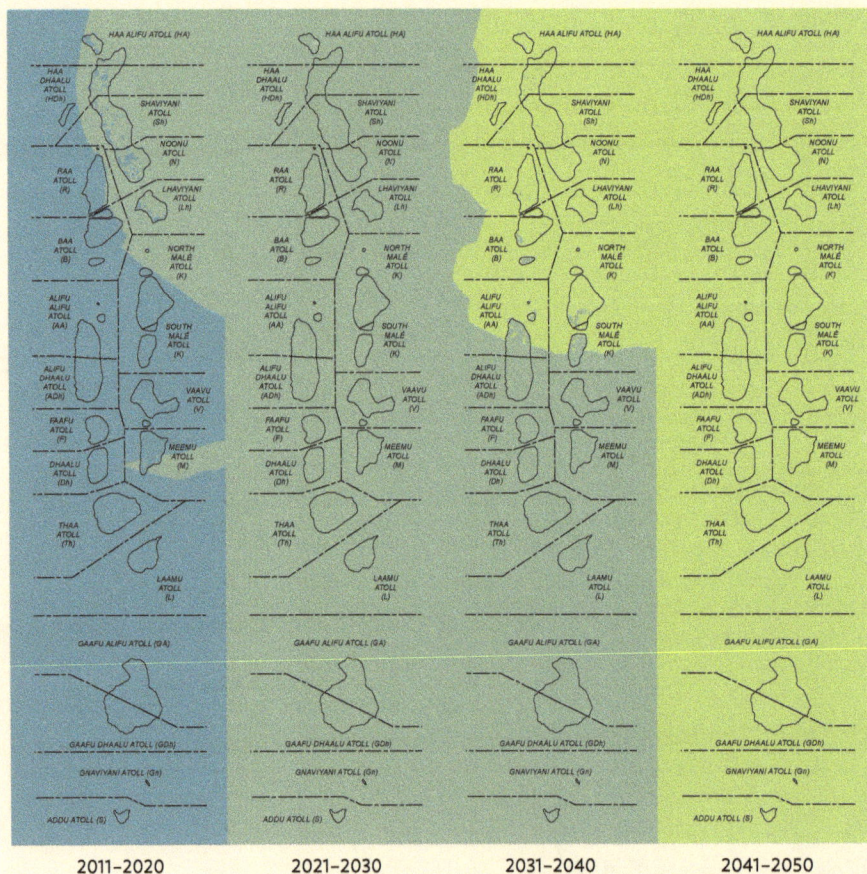

Legend

— Administrative Area
☐ Administrative Atoll

Temperature (°C)

27.40–27.95	30.15–30.70
27.95–28.50	30.70–31.25
28.50–29.05	31.25–31.80
29.05–29.60	31.80–32.35
29.60–30.15	32.35–32.95

Data Sources:
BODC, IHO, and IOC. 2003. GEBCO Digital Atlas (bathymetry); ME (administrative areas and atoll); and SANDER + PARTNER. 2017. www.sander-partner.com (temperature).

2011–2020   2021–2030   2031–2040   2041–2050

Average Seasonal Temperature Projection (MAM, RCP 4.5)

MAM are the warmest months, especially in northern Maldives. The average temperature ranges from 28.50°C to 29.05°C in the 2020s, warming to 29.60°C–30.70°C in the 2030s and 2040s.

☑	**Warmer north**
☑	**Warmer Maldives in the future**
☑	**Warmest months are MAM**

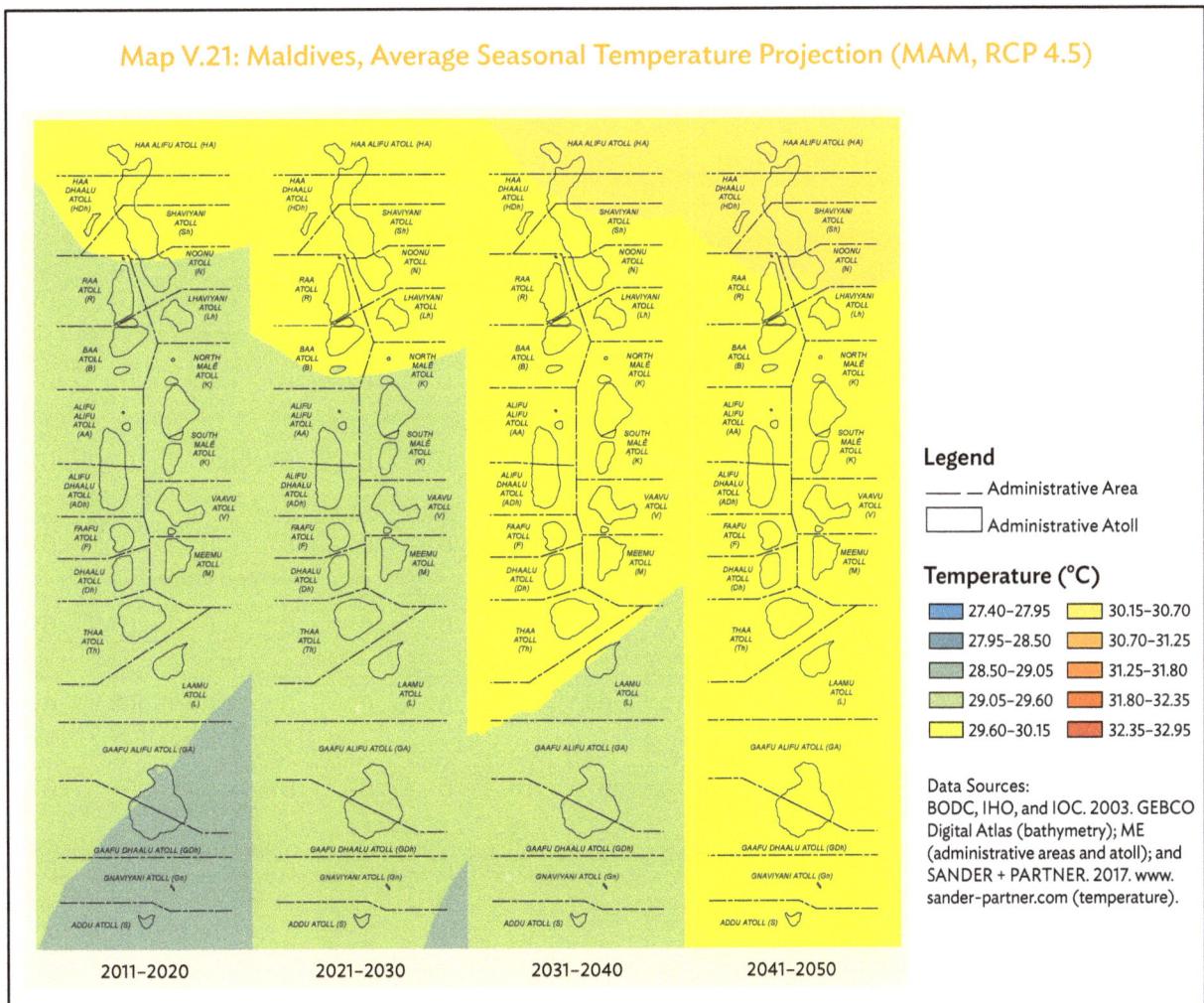

Map V.21: Maldives, Average Seasonal Temperature Projection (MAM, RCP 4.5)

Legend

— Administrative Area

☐ Administrative Atoll

Temperature (°C)

■	27.40–27.95	■	30.15–30.70
■	27.95–28.50	■	30.70–31.25
■	28.50–29.05	■	31.25–31.80
■	29.05–29.60	■	31.80–32.35
■	29.60–30.15	■	32.35–32.95

Data Sources:
BODC, IHO, and IOC. 2003. GEBCO Digital Atlas (bathymetry); ME (administrative areas and atoll); and SANDER + PARTNER. 2017. www.sander-partner.com (temperature).

2011–2020   2021–2030   2031–2040   2041–2050

Average Seasonal Temperature Projection (JJA, RCP 4.5)

☑ **Warmer north**
☑ **Warmer Maldives in the future**

The projected average temperature during JJA shows warming in the future, with the temperature rising from 27.95°C–28.50°C to 29.05°C–29.60°C beginning in 2031, especially in the northern and middle atolls.

Map V.22: Maldives, Average Seasonal Temperature Projection (JJA, RCP 4.5)

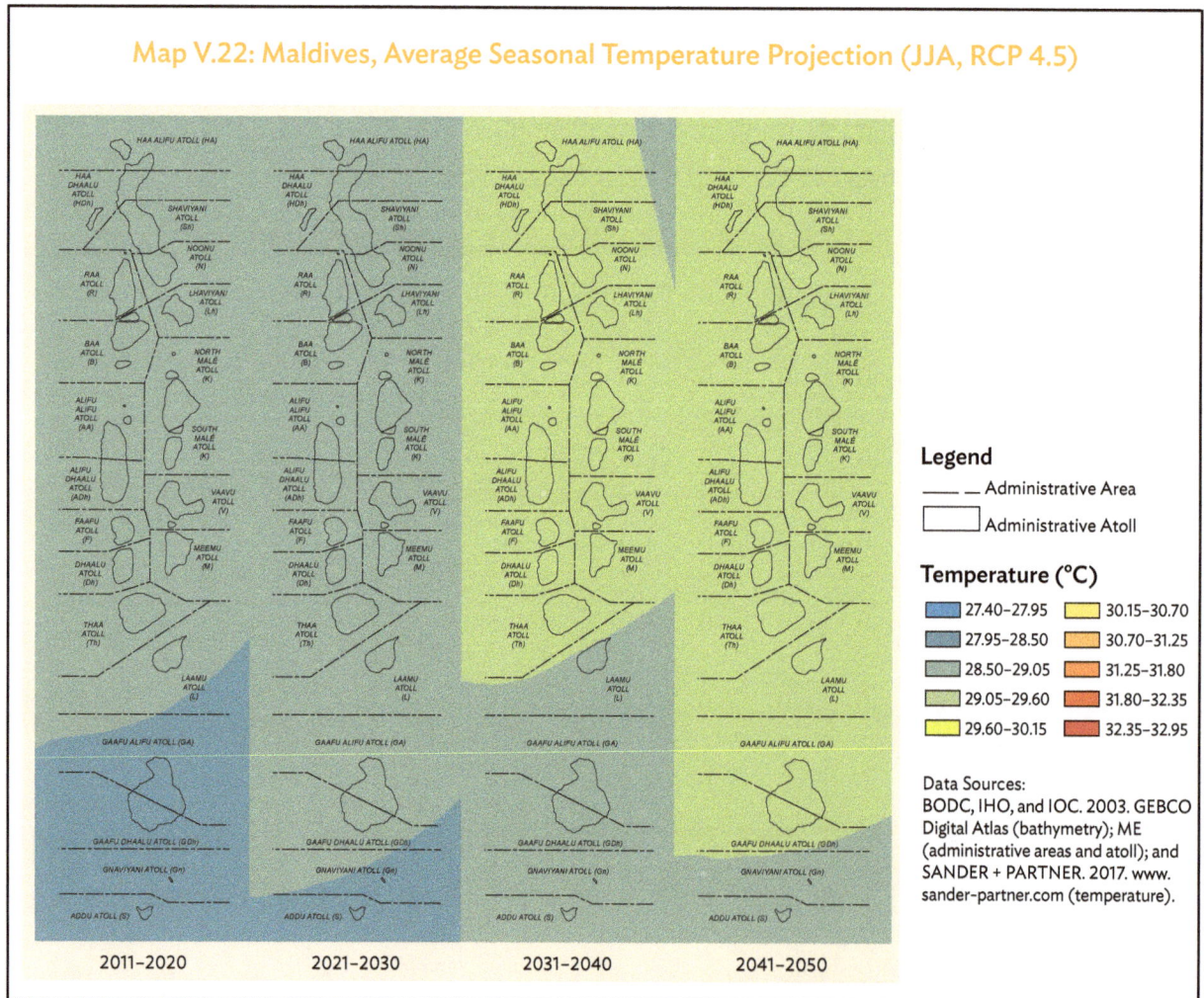

Legend

— Administrative Area
☐ Administrative Atoll

Temperature (°C)

27.40–27.95	30.15–30.70
27.95–28.50	30.70–31.25
28.50–29.05	31.25–31.80
29.05–29.60	31.80–32.35
29.60–30.15	32.35–32.95

Data Sources:
BODC, IHO, and IOC. 2003. GEBCO Digital Atlas (bathymetry); ME (administrative areas and atoll); and SANDER + PARTNER. 2017. www.sander-partner.com (temperature).

2011–2020   2021–2030   2031–2040   2041–2050

Average Seasonal Temperature Projection (SON, RCP 4.5)

SON are historically the coldest months of the year. The projected average temperature for SON shows they will continue to be Maldives' coldest months. However, there is a general increase in the average temperature from 27.95°C–28.50°C in 2011–2020 to 28.50°C–29.05°C in 2041–2050.

- ☑ **Warmer north**
- ☑ **Warmer Maldives in the future**
- ☑ **SON are the coldest months**

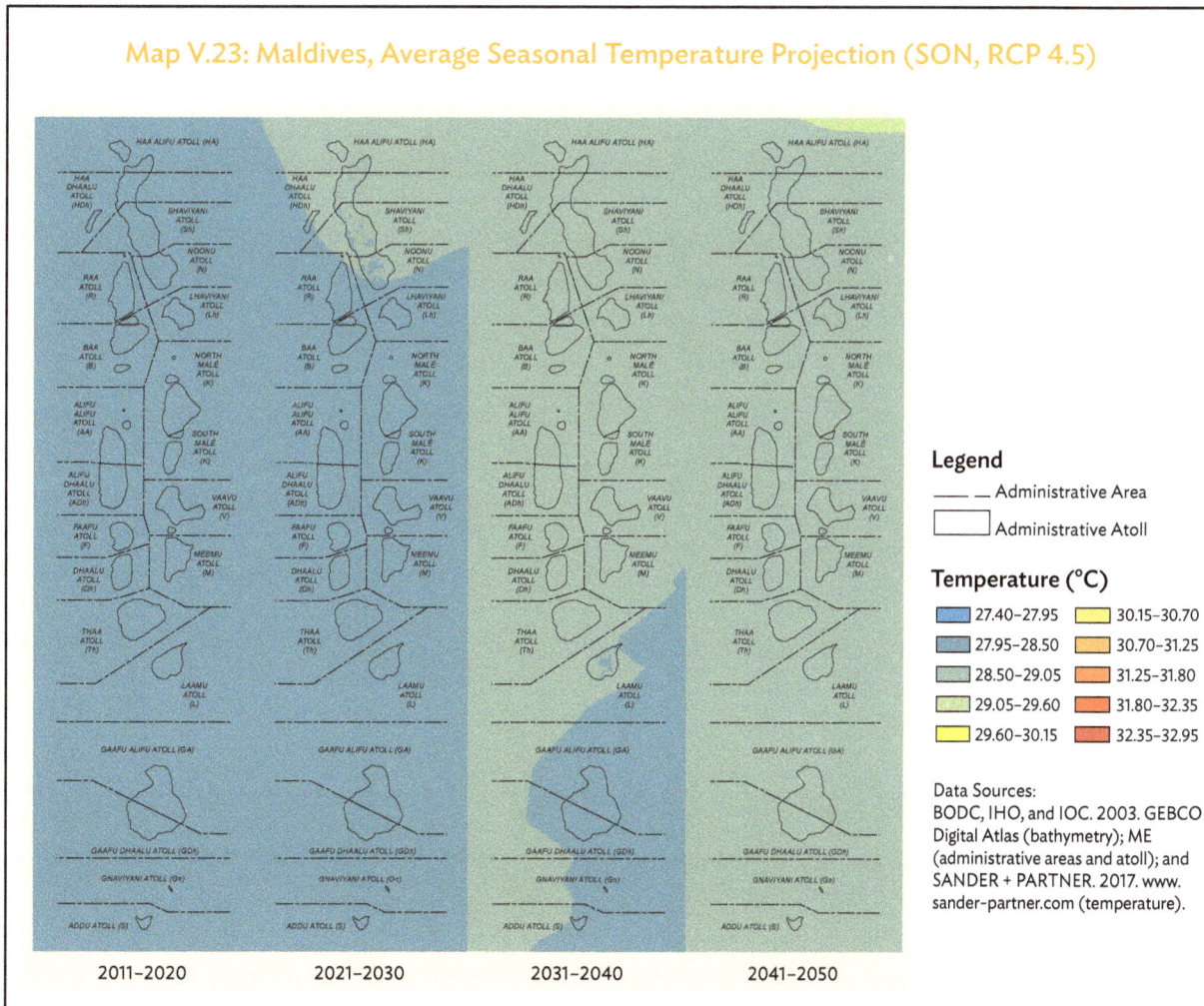

Map V.23: Maldives, Average Seasonal Temperature Projection (SON, RCP 4.5)

Legend

—— Administrative Area

☐ Administrative Atoll

Temperature (°C)

27.40–27.95	30.15–30.70
27.95–28.50	30.70–31.25
28.50–29.05	31.25–31.80
29.05–29.60	31.80–32.35
29.60–30.15	32.35–32.95

Data Sources:
BODC, IHO, and IOC. 2003. GEBCO Digital Atlas (bathymetry); ME (administrative areas and atoll); and SANDER + PARTNER. 2017. www.sander-partner.com (temperature).

2011–2020 2021–2030 2031–2040 2041–2050

Average Annual Temperature Projection (RCP 8.5)

- ☑ **Warmer north**
- ☑ **Warmer Maldives in the future**

In the RCP 8.5 scenario, Maldives is projected to be warmer in the 2030s and 2040s, especially in the northern atolls. Average temperature will increase from 29.05°C–29.60°C in the 2020s and 2030s to 29.60°C–30.15°C in the 2040s.

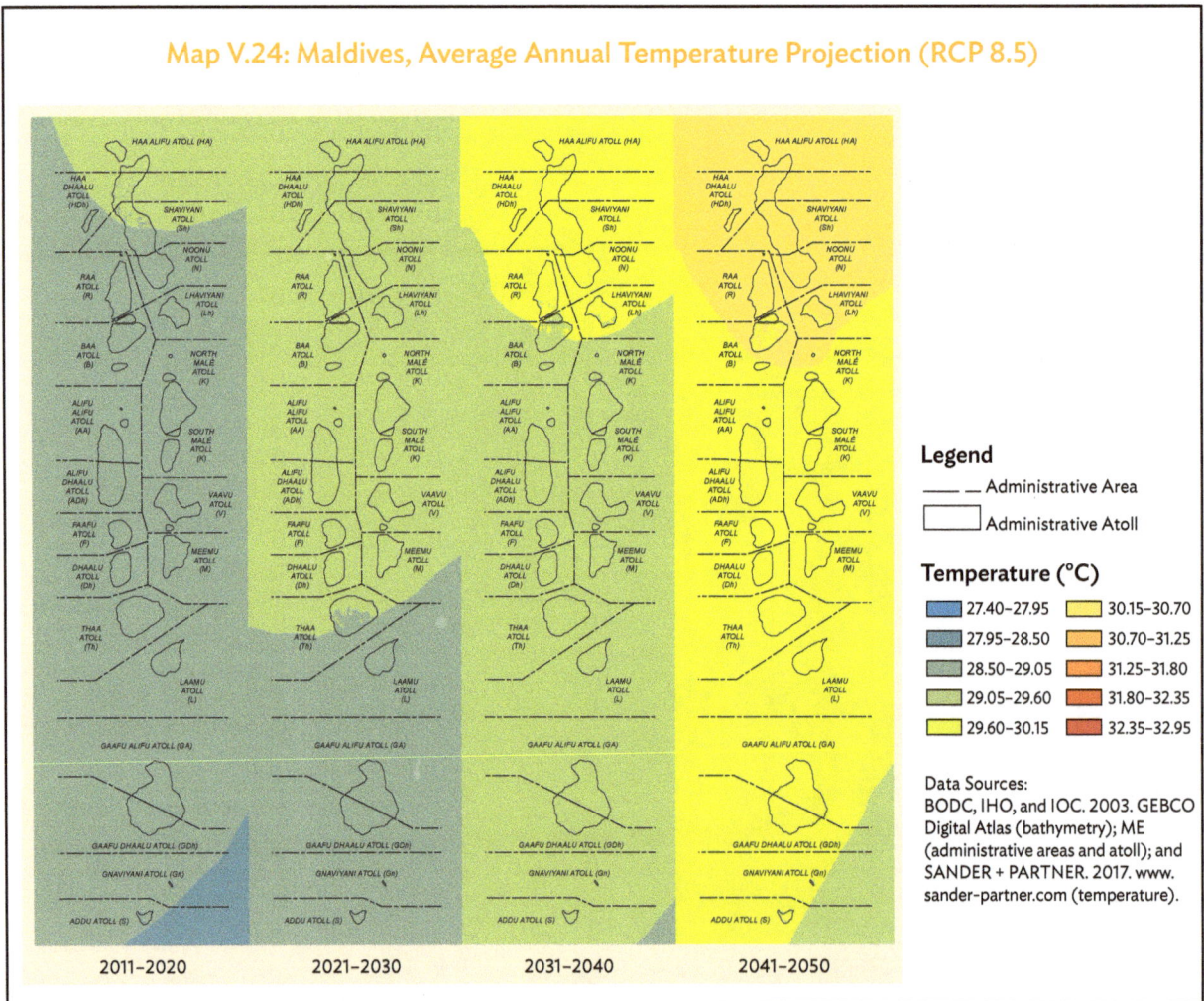

Map V.24: Maldives, Average Annual Temperature Projection (RCP 8.5)

Legend
— Administrative Area
☐ Administrative Atoll

Temperature (°C)

27.40–27.95	30.15–30.70
27.95–28.50	30.70–31.25
28.50–29.05	31.25–31.80
29.05–29.60	31.80–32.35
29.60–30.15	32.35–32.95

Data Sources:
BODC, IHO, and IOC. 2003. GEBCO Digital Atlas (bathymetry); ME (administrative areas and atoll); and SANDER + PARTNER. 2017. www.sander-partner.com (temperature).

2011–2020 · 2021–2030 · 2031–2040 · 2041–2050

Average Seasonal Temperature Projection (DJF, RCP 8.5)

The RCP 8.5 average temperature projection for DJF (average range of 28.50°C–29.05°C) shows no spatial variation in the temperature across Maldives from 2011 to 2030. The temperature will start to rise in the 2030s, reaching 29.05°C–29.60°C. By the 2040s, the northern and middle portion of Maldives will be warmer, with an average range of 29.60°C–30.15°C.

- ☑ **Warmer north**
- ☑ **Warming in northern Maldives in the future**
- ☑ **RCP 8.5 projected higher temperature than RCP 4.5**

Map V.25: Maldives, Average Seasonal Temperature Projection (DJF, RCP 8.5)

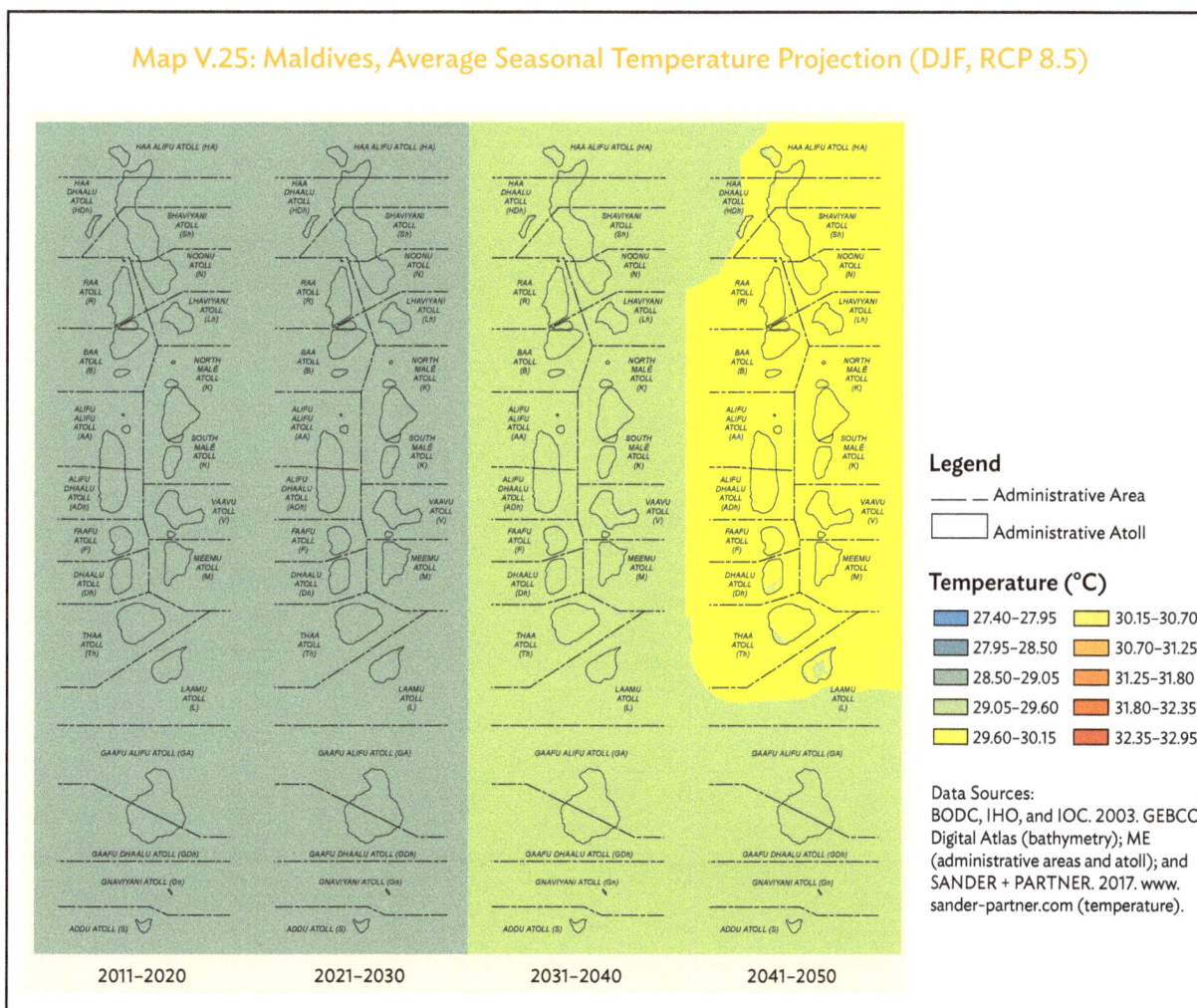

Legend

— Administrative Area
☐ Administrative Atoll

Temperature (°C)

27.40–27.95	30.15–30.70
27.95–28.50	30.70–31.25
28.50–29.05	31.25–31.80
29.05–29.60	31.80–32.35
29.60–30.15	32.35–32.95

Data Sources:
BODC, IHO, and IOC. 2003. GEBCO Digital Atlas (bathymetry); ME (administrative areas and atoll); and SANDER + PARTNER. 2017. www.sander-partner.com (temperature).

2011–2020 | 2021–2030 | 2031–2040 | 2041–2050

Average Seasonal Temperature Projection (MAM, RCP 8.5)

- ☑ **Warmer north**
- ☑ **Warmer Maldives in the future**

The RCP 8.5 average temperature projection shows that MAM in Maldives will be warmer at 29.60°C–30.15°C in the 2020s and 2030s. Northern Maldives will experience warmer temperature (reaching 30.15°C–30.70°C in the 2040s) during these months.

Map V.26: Maldives, Average Seasonal Temperature Projection (MAM, RCP 8.5)

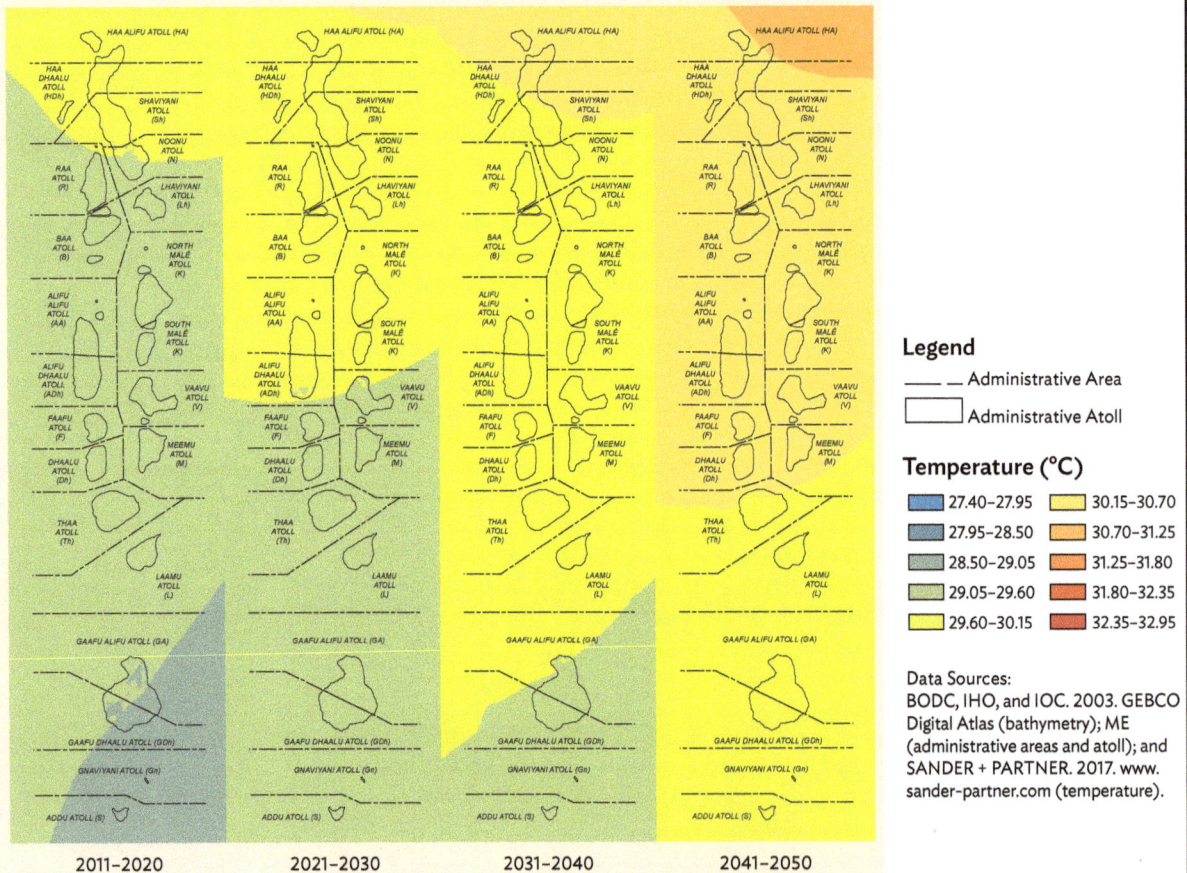

Legend

— — Administrative Area

☐ Administrative Atoll

Temperature (°C)

27.40–27.95	30.15–30.70
27.95–28.50	30.70–31.25
28.50–29.05	31.25–31.80
29.05–29.60	31.80–32.35
29.60–30.15	32.35–32.95

Data Sources:
BODC, IHO, and IOC. 2003. GEBCO Digital Atlas (bathymetry); ME (administrative areas and atoll); and SANDER + PARTNER. 2017. www.sander-partner.com (temperature).

2011–2020   2021–2030   2031–2040   2041–2050

Average Seasonal Temperature Projection (JJA, RCP 8.5)

For JJA, the pattern of average seasonal projected temperature based on RCP 8.5 shows that the northern and middle portions of Maldives are warmer compared to the southern region. JJA will be warmer in the coming decades, especially in the 2040s, as atolls experience temperatures ranging from 29.05°C to 30.15°C, with higher temperatures in the middle portion of Maldives. As with other seasonal projections, RCP 8.5 depicts a warmer scenario compared to RCP 4.5.

- ☑ **Warmer north**
- ☑ **Warming in middle Maldives in the future**
- ☑ **RCP 8.5 projected higher temperature in 2041–2050 than RCP 4.5**

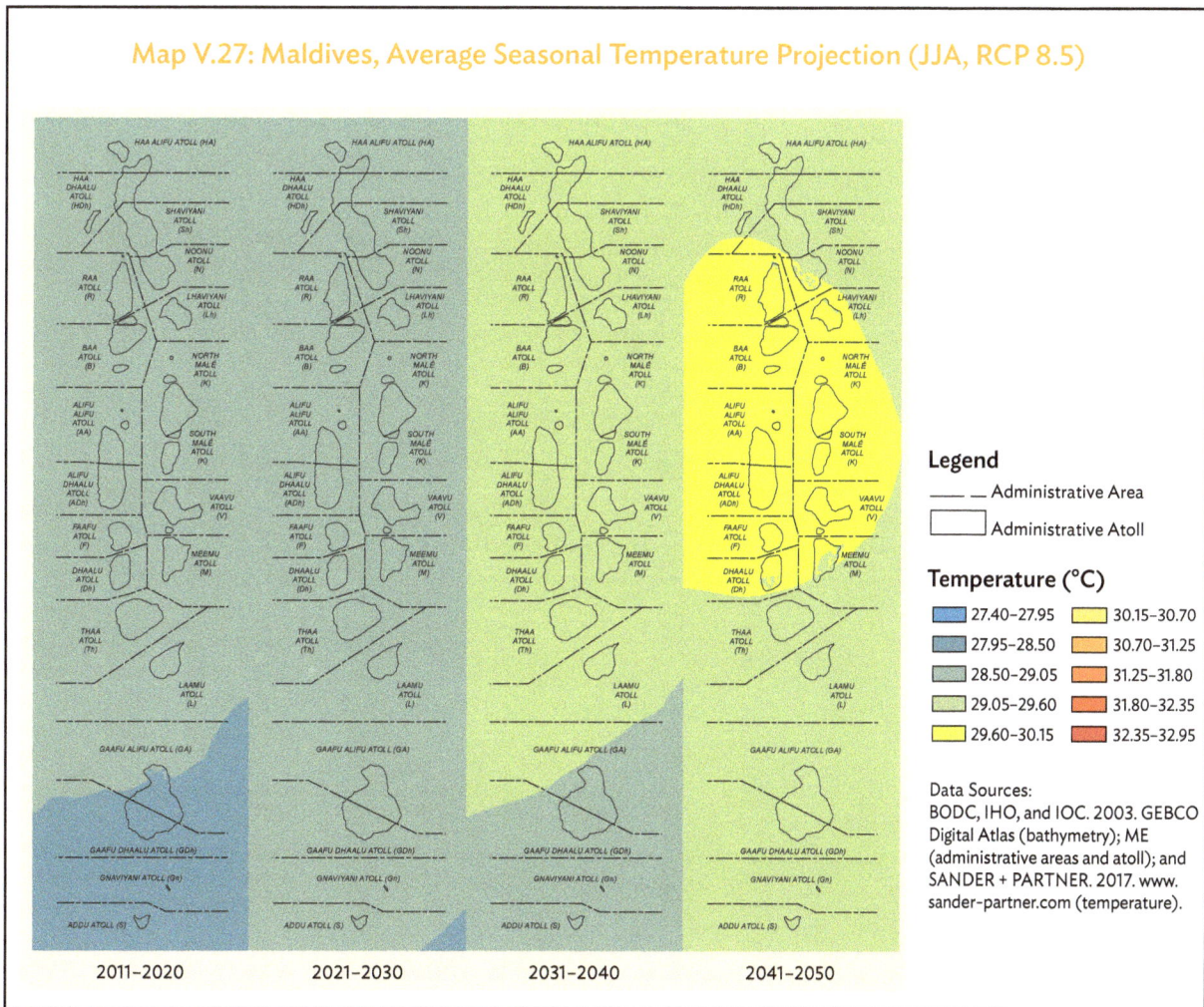

Map V.27: Maldives, Average Seasonal Temperature Projection (JJA, RCP 8.5)

Legend

— Administrative Area

☐ Administrative Atoll

Temperature (°C)

27.40–27.95	30.15–30.70
27.95–28.50	30.70–31.25
28.50–29.05	31.25–31.80
29.05–29.60	31.80–32.35
29.60–30.15	32.35–32.95

Data Sources:
BODC, IHO, and IOC. 2003. GEBCO Digital Atlas (bathymetry); ME (administrative areas and atoll); and SANDER + PARTNER. 2017. www.sander-partner.com (temperature).

2011–2020   2021–2030   2031–2040   2041–2050

☑ **Warmer north (2020s and 2040s)**

☑ **Warming in the future**

☑ **RCP 8.5 projected higher temperature in 2041–2050 than RCP 4.5**

SON will still be the coldest months, according to the average temperature projection using RCP 8.5. The northern half of Maldives will be warmer in the 2020s and 2040s, compared to the 2010s and 2030s when the temperature is even from north to south. The coming decades will be warmer as the temperature increases from 27.95°C–28.50°C in the 2010s to 29.05°C–29.60°C in the 2040s.

Map V.28: Maldives, Average Seasonal Temperature Projection (SON, RCP 8.5)

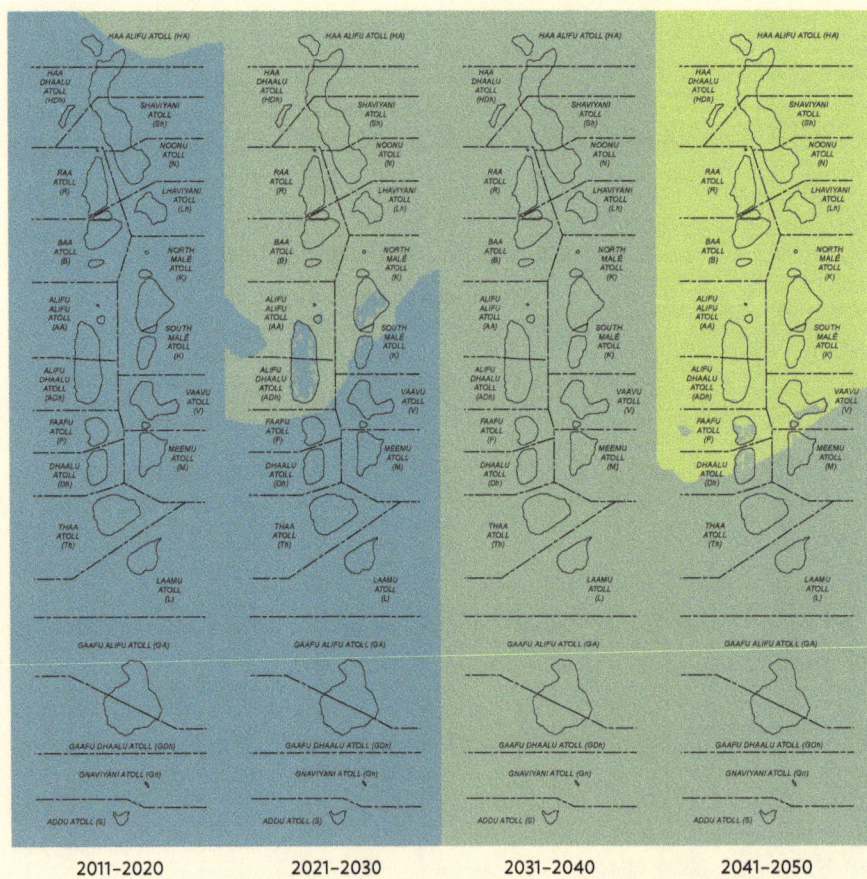

Legend

— Administrative Area

☐ Administrative Atoll

Temperature (°C)

27.40–27.95	30.15–30.70
27.95–28.50	30.70–31.25
28.50–29.05	31.25–31.80
29.05–29.60	31.80–32.35
29.60–30.15	32.35–32.95

Data Sources:
BODC, IHO, and IOC. 2003. GEBCO Digital Atlas (bathymetry); ME (administrative areas and atoll); and SANDER + PARTNER. 2017. www.sander-partner.com (temperature).

2011–2020   2021–2030   2031–2040   2041–2050

Summary of Observations for Rainfall

Similar to the historical annual average rainfall distribution, northern Maldives will continue to receive more rainfall compared to the southern region. However, this may vary depending on the season. The wettest months will still be JJA. During this period, rainfall will mostly be distributed in the north. Except for the slight variation in the area receiving 4–8 mm/day of rainfall, the projections for RCP 4.5 and RCP 8.5 show no difference.

Table V.4: Variation in Rainfall Patterns across Space, Time, and Representative Concentration Pathways

Climate Maps	Spatial	Seasonal	Decadal	RCP 4.5 vs RCP 8.5
Average Annual	Wetter north (4.6–5.8 mm/day)	-	-	-
Average Seasonal	Wetter north	Generally dry season	-	-
Projected Mean Annual (4.5)	Wetter north	-	Wetter north in the future	No difference with 8.5
Projected Mean Seasonal (4.5) DJF	Wetter south	Dry season	Wetter southern atolls	Smaller coverage of 4–8 mm/day
Projected Mean Seasonal (4.5) MAM	Generally dry	Driest season	No change	No difference with 8.5
Projected Mean Seasonal (4.5) JJA	Wetter north	Wettest season	Wetter north in the future	No difference with 8.5
Projected Mean Seasonal (4.5) SON	Wetter north	Wet season	No change	Smaller coverage of 4–8 mm/day
Projected Mean Annual (8.5)	Wetter north	-	Wetter north in the future	No difference with 4.5
Projected Mean Seasonal (8.5) DJF	Wetter south	Dry season	More atolls in the south will experience higher rainfall rate	Greater coverage of 4–8 mm/day
Projected Mean Seasonal (8.5) MAM	Generally dry	Driest season	No change	No difference with 4.5
Projected Mean Seasonal (8.5) JJA	Wetter north	Wettest season	Wetter north in the future	No difference with 4.5
Projected Mean Seasonal (8.5) SON	Wetter north	Wet season	More atolls in the north with 4–8 mm/day rainfall rate	Greater coverage of 4–8 mm/day

– = not applicable; DJF = December, January, February; JJA = June, July, August; MAM = March, April, May; mm/day = millimeter per day; RCP = representative concentration pathway; SON = September, October, November.

Source: SANDER + PARTNER. 2017. www.sander-partner.com.

Summary of Observations for Temperature

Maldives will experience a warmer climate in the 2030s and 2040s. Greater warming will be felt in the northern atolls, especially in MAM. This pattern is consistent across seasons and decades. However, the RCP 8.5 projections show a significantly warmer possibility compared with the RCP 4.5.

Table V.5: Variation in Temperature Patterns across Space, Time, and Representative Concentration Pathways

Climate Maps	Spatial	Seasonal	Decadal	RCP 4.5 vs RCP 8.5
Average Annual	Warmer north (28–28.4°C)	–	Warmer Maldives especially in the future	–
Average Seasonal	–	MAM is the warmest month, SON is the coldest month	–	–
Projected Mean Annual (4.5)	Warmer north	–	Future warmer Maldives especially in the north	Lower temperature compared to 8.5
Projected Mean Seasonal (4.5) DJF	Warmer north	–	Warmer Maldives especially in the future	Lower temperature compared to 8.5
Projected Mean Seasonal (4.5) MAM	Warmer north	MAM is the warmest month	Future warmer Maldives especially in the north	Lower temperature compared to 8.5
Projected Mean Seasonal (4.5) JJA	Warmer north	–	Future warmer Maldives especially in the north	Lower temperature compared to 8.5
Projected Mean Seasonal (4.5) SON	Warmer north	SON is the coldest month	Future warmer Maldives especially in the north	Lower temperature compared to 8.5
Projected Mean Annual (8.5)	Warmer north	–	Future warmer Maldives especially in the north	Higher temperature compared to 4.5
Projected Mean Seasonal (8.5) DJF	Even temperature across Maldives except in 2040s	–	Future warmer Maldives especially in the north	Higher temperature compared to 4.5
Projected Mean Seasonal (8.5) MAM	Warmer north	Warmest months	Future warmer Maldives especially in the north	Higher temperature compared to 4.5
Projected Mean Seasonal (8.5) JJA	Warmer north	–	Warming in the middle of Maldives	Higher temperature compared to 4.5
Projected Mean Seasonal (8.5) SON	Warmer north in 2020s and 2040s	Coldest months	Future warmer Maldives especially in the north	Higher temperature compared to 4.5

– = not applicable; °C = degree Celsius; DJF = December, January, February; JJA = June, July, August; MAM = March, April, May; RCP = representative concentration pathway; SON = September, October, November.

Source: SANDER + PARTNER. 2017. www.sander-partner.com.

Geophysical Hazards

Maldives has experienced several disasters in the past. According to a 2006 United Nations Development Programme (UNDP) report in which cyclone tracks were analyzed across the century, Maldives has only experienced 11 cyclones in 12 decades, with the strongest cyclones crossing northern Maldives (UNDP 2006). Although cyclones are not a frequently recurring hazard in the country, they still bring wind, rain, and storm surges to the low-lying islands. Flooding, usually caused by a surge, has recurred several times in Maldives. *Bodu raalhu* (big waves) flooded a few islands in 1987. Northeastern islands are mostly affected by storm surges (UNDP 2006).

Aside from cyclones, Maldives also experiences earthquakes and tsunamis. In 25 years, three earthquakes with a magnitude of at least 7.0 hit Maldives (UNDP 2006). The UNDP report also estimated the decay of peak ground acceleration for a 475-year return period. Results showed that the southern Gnaviyani Atoll and Addu Atoll have the highest peak ground acceleration. Movements at subduction zones, usually occurring at the edges of the tectonic plates, generate earthquakes, which in turn generate tsunamis. The most devastating disaster was the tsunami on 26 December 2004. It generated waves up to 4.2 meters that damaged at least a dozen of the inhabited islands and thousands of homes, affected thousands of lives, and lost about two-thirds of the country's gross domestic product (UNDP 2006). Based on UNDP's model, the eastern borders of atolls have higher probability of experiencing tsunamis at 320–450 centimeters high.

Tsunami Monument. Various hazards, including tsunamis, have affected Maldives in the past. This Tsunami Monument stands at the shores of Malé in memory of the people who died during the 2004 tsunami (photo by Utsav Mulay).

Map V.29: Maldives, Cyclonic Wind Hazard Zone

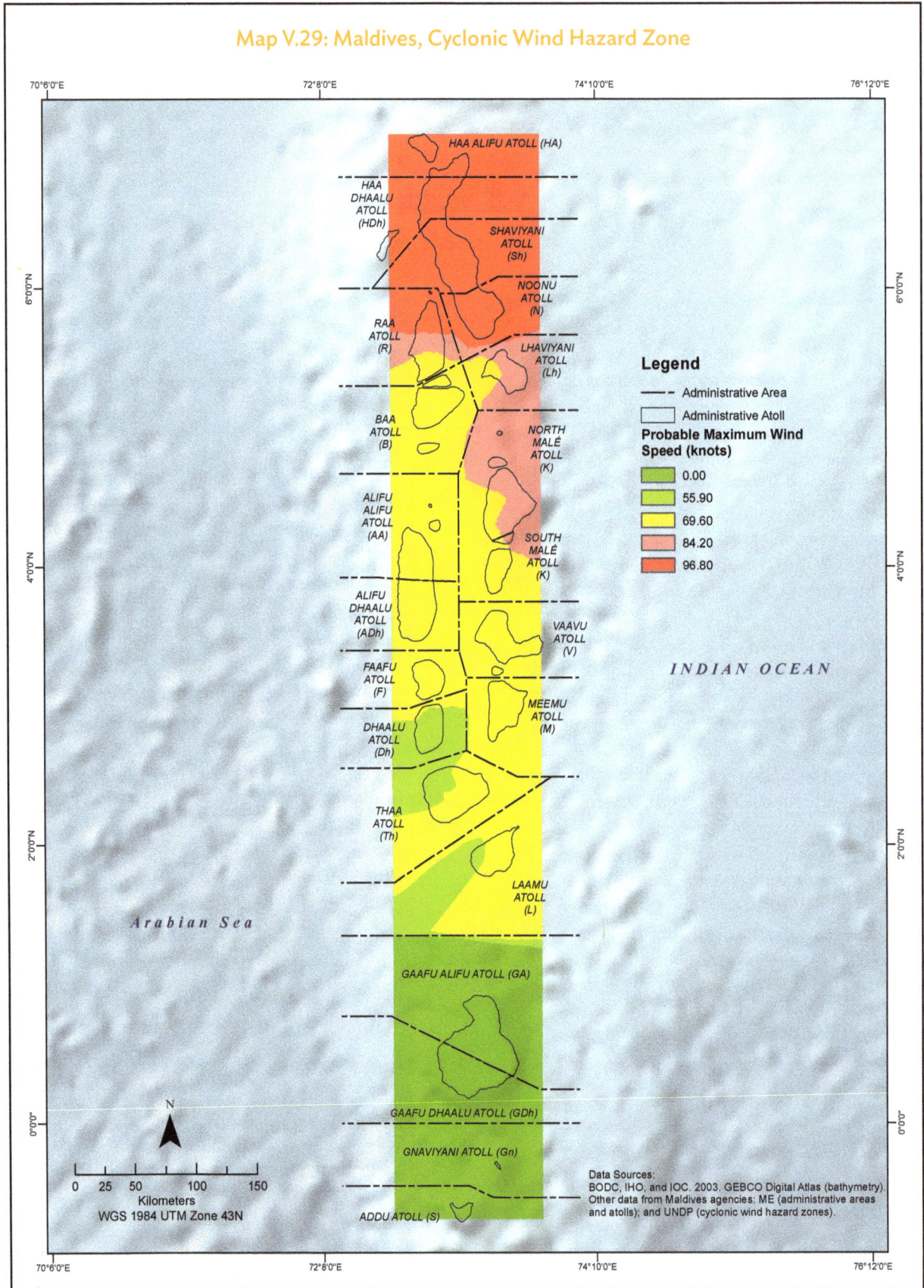

Legend

- – – – Administrative Area
- ▭ Administrative Atoll

Probable Maximum Wind Speed (knots)

- 0.00
- 55.90
- 69.60
- 84.20
- 96.80

HAA ALIFU ATOLL (HA)

HAA DHAALU ATOLL (HDh)

SHAVIYANI ATOLL (Sh)

NOONU ATOLL (N)

RAA ATOLL (R)

LHAVIYANI ATOLL (Lh)

BAA ATOLL (B)

NORTH MALÉ ATOLL (K)

ALIFU ALIFU ATOLL (AA)

SOUTH MALÉ ATOLL (K)

ALIFU DHAALU ATOLL (ADh)

VAAVU ATOLL (V)

FAAFU ATOLL (F)

MEEMU ATOLL (M)

DHAALU ATOLL (Dh)

THAA ATOLL (Th)

LAAMU ATOLL (L)

GAAFU ALIFU ATOLL (GA)

GAAFU DHAALU ATOLL (GDh)

GNAVIYANI ATOLL (Gn)

ADDU ATOLL (S)

INDIAN OCEAN

Arabian Sea

N

0 25 50 100 150
Kilometers
WGS 1984 UTM Zone 43N

Data Sources:
BODC, IHO, and IOC. 2003. GEBCO Digital Atlas (bathymetry).
Other data from Maldives agencies: ME (administrative areas
and atolls); and UNDP (cyclonic wind hazard zones).

Map V.30: Maldives, Surge Hazard Zone

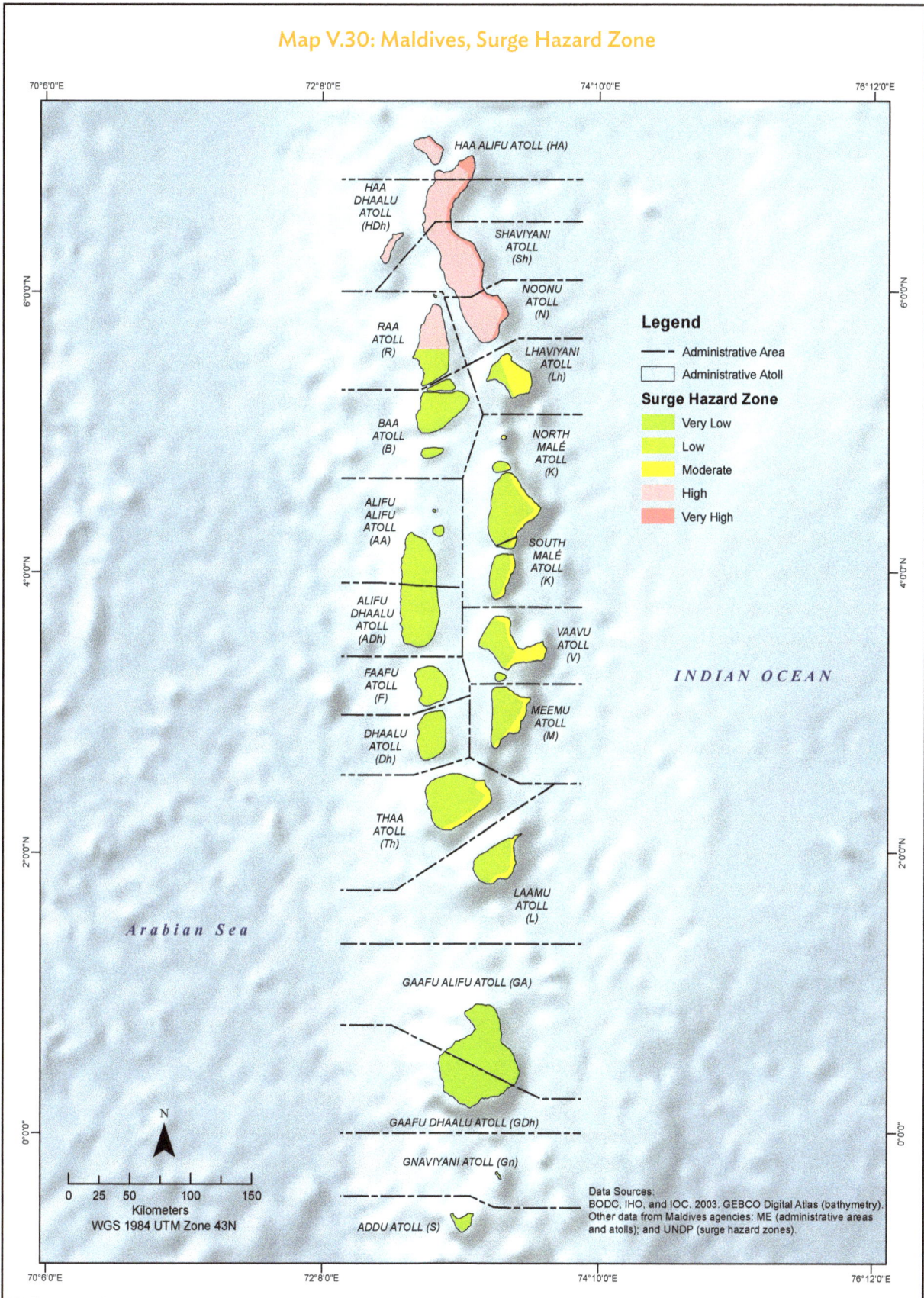

HAA ALIFU ATOLL (HA)

HAA DHAALU ATOLL (HDh)

SHAVIYANI ATOLL (Sh)

NOONU ATOLL (N)

RAA ATOLL (R)

LHAVIYANI ATOLL (Lh)

BAA ATOLL (B)

NORTH MALÉ ATOLL (K)

ALIFU ALIFU ATOLL (AA)

SOUTH MALÉ ATOLL (K)

ALIFU DHAALU ATOLL (ADh)

VAAVU ATOLL (V)

FAAFU ATOLL (F)

MEEMU ATOLL (M)

DHAALU ATOLL (Dh)

THAA ATOLL (Th)

LAAMU ATOLL (L)

GAAFU ALIFU ATOLL (GA)

GAAFU DHAALU ATOLL (GDh)

GNAVIYANI ATOLL (Gn)

ADDU ATOLL (S)

INDIAN OCEAN

Arabian Sea

Legend

— · — Administrative Area

☐ Administrative Atoll

Surge Hazard Zone

- Very Low
- Low
- Moderate
- High
- Very High

N

0 25 50 100 150
Kilometers
WGS 1984 UTM Zone 43N

Data Sources:
BODC, IHO, and IOC. 2003. GEBCO Digital Atlas (bathymetry).
Other data from Maldives agencies: ME (administrative areas and atolls); and UNDP (surge hazard zones).

70°6'0"E   72°8'0"E   74°10'0"E   76°12'0"E

6°0'0"N   4°0'0"N   2°0'0"N   0°0'0"

Map V.31: Maldives, Seismic Hazard Zone

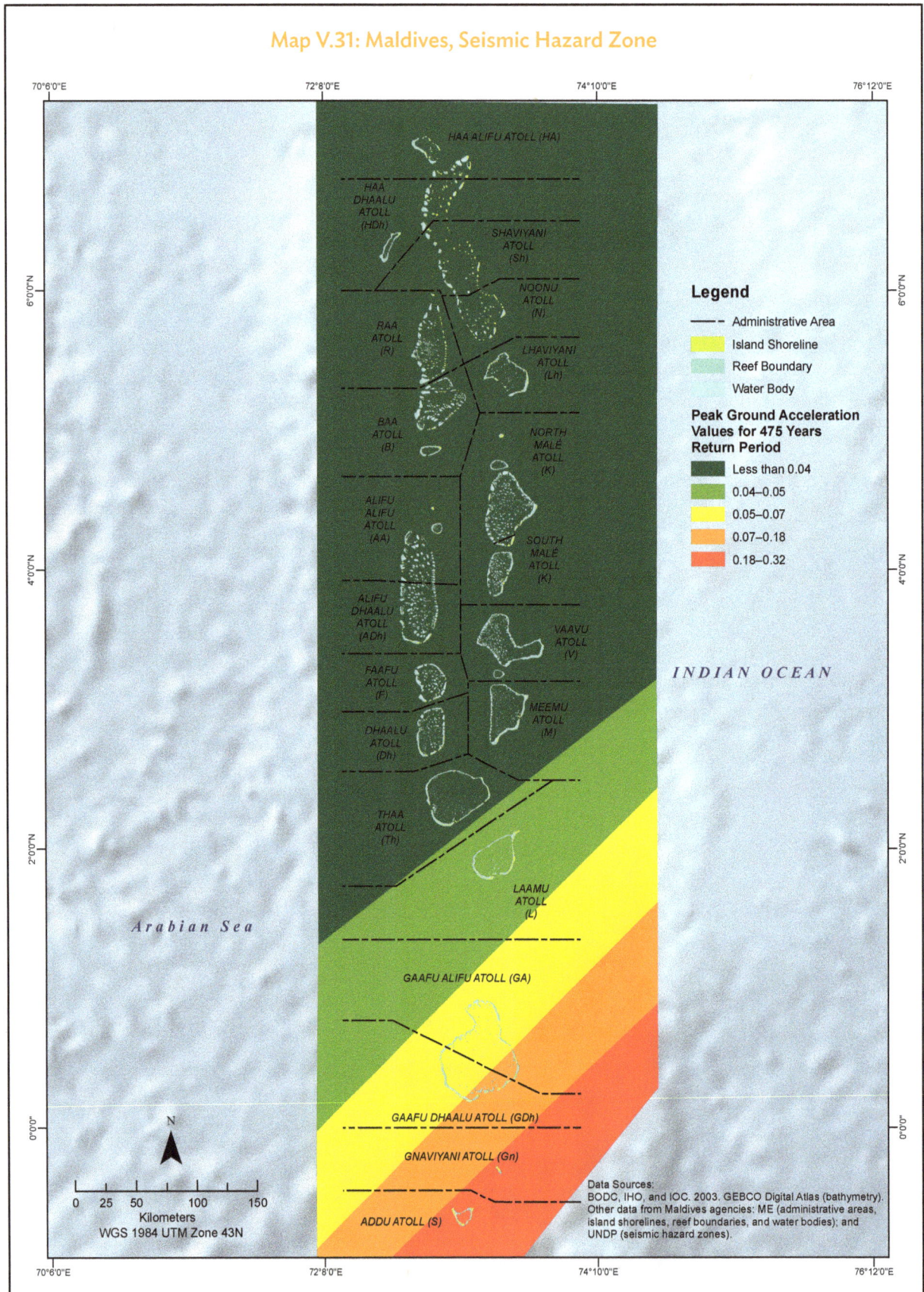

Legend

- – – Administrative Area
- Island Shoreline
- Reef Boundary
- Water Body

Peak Ground Acceleration Values for 475 Years Return Period

- Less than 0.04
- 0.04–0.05
- 0.05–0.07
- 0.07–0.18
- 0.18–0.32

HAA ALIFU ATOLL (HA)
HAA DHAALU ATOLL (HDh)
SHAVIYANI ATOLL (Sh)
NOONU ATOLL (N)
RAA ATOLL (R)
LHAVIYANI ATOLL (Lh)
BAA ATOLL (B)
NORTH MALÉ ATOLL (K)
ALIFU ALIFU ATOLL (AA)
SOUTH MALÉ ATOLL (K)
ALIFU DHAALU ATOLL (ADh)
VAAVU ATOLL (V)
FAAFU ATOLL (F)
DHAALU ATOLL (Dh)
MEEMU ATOLL (M)
THAA ATOLL (Th)
LAAMU ATOLL (L)
GAAFU ALIFU ATOLL (GA)
GAAFU DHAALU ATOLL (GDh)
GNAVIYANI ATOLL (Gn)
ADDU ATOLL (S)

INDIAN OCEAN

Arabian Sea

N

0 25 50 100 150
Kilometers
WGS 1984 UTM Zone 43N

Data Sources:
BODC, IHO, and IOC. 2003. GEBCO Digital Atlas (bathymetry). Other data from Maldives agencies: ME (administrative areas, island shorelines, reef boundaries, and water bodies); and UNDP (seismic hazard zones).

Map V.32: Maldives, Tsunami Hazard Zone

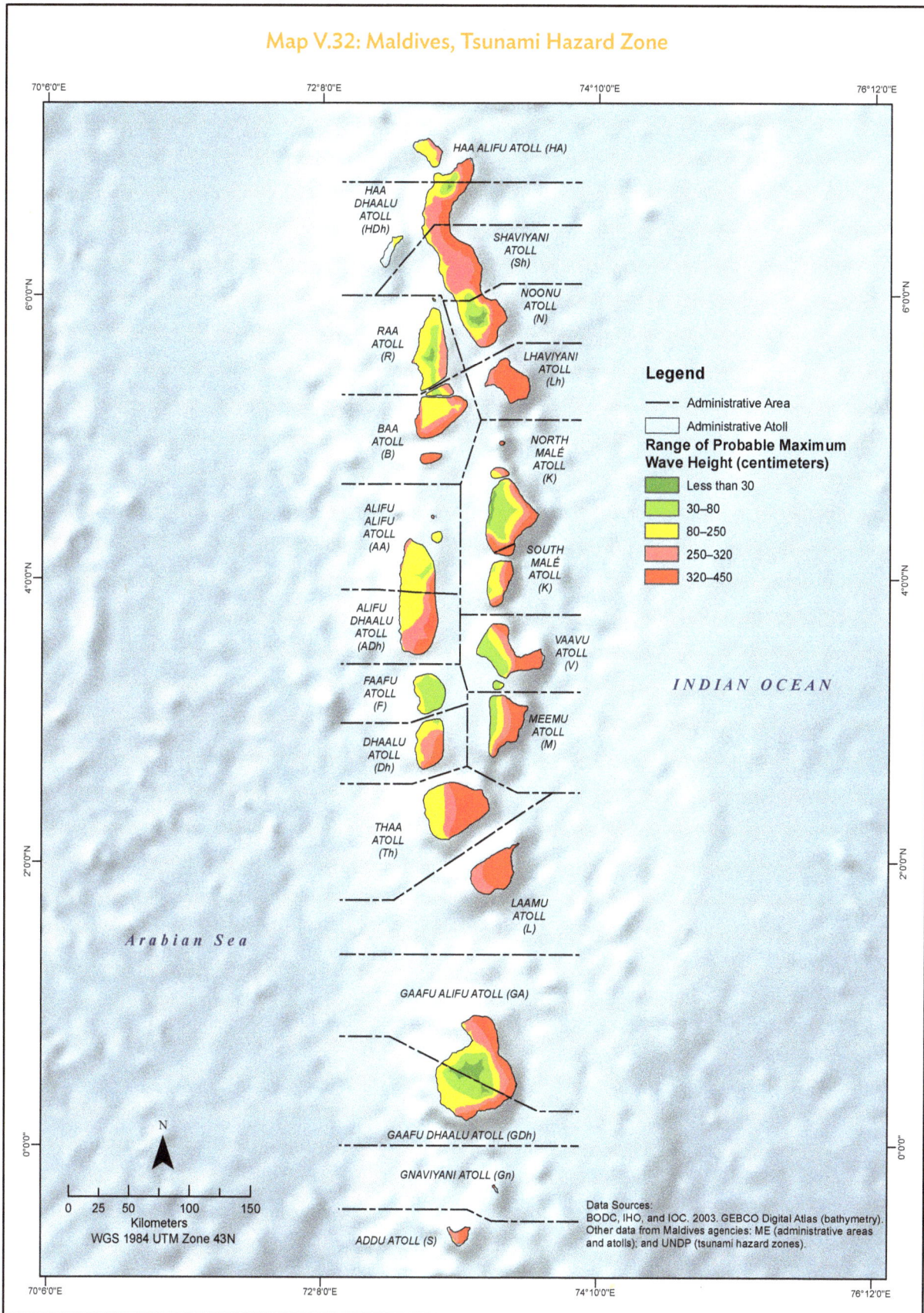

Legend

- - - Administrative Area

☐ Administrative Atoll

Range of Probable Maximum Wave Height (centimeters)

- Less than 30
- 30–80
- 80–250
- 250–320
- 320–450

INDIAN OCEAN

Arabian Sea

N

| 0 25 50   100   150 |
Kilometers
WGS 1984 UTM Zone 43N

Data Sources:
BODC, IHO, and IOC. 2003. GEBCO Digital Atlas (bathymetry).
Other data from Maldives agencies: ME (administrative areas
and atolls); and UNDP (tsunami hazard zones).

HAA ALIFU ATOLL (HA)
HAA DHAALU ATOLL (HDh)
SHAVIYANI ATOLL (Sh)
NOONU ATOLL (N)
RAA ATOLL (R)
LHAVIYANI ATOLL (Lh)
BAA ATOLL (B)
NORTH MALÉ ATOLL (K)
ALIFU ALIFU ATOLL (AA)
SOUTH MALÉ ATOLL (K)
ALIFU DHAALU ATOLL (ADh)
VAAVU ATOLL (V)
FAAFU ATOLL (F)
MEEMU ATOLL (M)
DHAALU ATOLL (Dh)
THAA ATOLL (Th)
LAAMU ATOLL (L)
GAAFU ALIFU ATOLL (GA)
GAAFU DHAALU ATOLL (GDh)
GNAVIYANI ATOLL (Gn)
ADDU ATOLL (S)

Demographics and Economy

Humans of Maldives

This section examines Maldivians by focusing on the spatial characteristics of the demographics and economic activities in the country. It concentrates on the people at risk to climate and geologic hazards. This population runs the country—and its economy and politics—and cultivates a unique blend of culture.

Specifically, this section looks at the distribution of the population across the islands; the allocation of educational and health facilities, tourist spots, transportation and other infrastructure; and the facilities and locations of economic activities, particularly sand mining. All these elements have varying exposure and vulnerability to climate and disaster risks.

Activities. People gather on the beach for the Maahefun Festival in Fuvahmulah (photo by Ibrahim Asad).

Fishing. Fishermen bring their catch to the butcher for cleaning in preparation for selling in the market (photo by Mark Fischer).

Population

Population fundamentally measures exposure to hazards. The number of people exposed to certain hazards are based on population density.

Maldives has the lowest population density among countries in Asia and the Pacific. More than 400,000 Maldivians are scattered across the roughly 200 inhabited islands in 21 atolls (National Bureau of Statistics 2014). Roughly 38% of the population resides in Malé City, the largest city in Maldives. The rest are distributed across the atolls.

Table V.6: Atoll Population in Maldives

Atoll	Population 2014	%
Malé City	153,904	38.28
Kaafu Atoll	27,360	6.80
Seenu Atoll	21,957	5.46
Haa Dhaalu Atoll	19,649	4.89
Raa Atoll	16,649	4.14
Haa Alifu Atoll	14,666	3.65
Baa Atoll	13,507	3.36
Laamu Atoll	13,498	3.36
Alifu Dhaalu Atoll	13,256	3.30
Gaafu Dhaalu Atoll	13,104	3.26
Shaviyani Atoll	13,095	3.26
Noonu Atoll	12,831	3.19
Gaafu Alifu Atoll	10,862	2.70
Lhaviyani Atoll	10,659	2.65
Thaa Atoll	9,893	2.46
Alifu Alifu Atoll	9,075	2.26
Gnaviyani Atoll	8,510	2.12
Dhaalu Atoll	7,448	1.85
Meemu Atoll	5,478	1.36
Faafu Atoll	4,627	1.15
Vaavu Atoll	2,043	0.51

Note: Total may not add up due to rounding.

Source: National Bureau of Statistics, 2014.

People of Maldives. People enjoy spending time in the beaches of Maldives, which are famous for their natural beauty, clear, turquoise waters, and fine white sand (photo by Adam Azim).

Map V.33: Maldives, Population

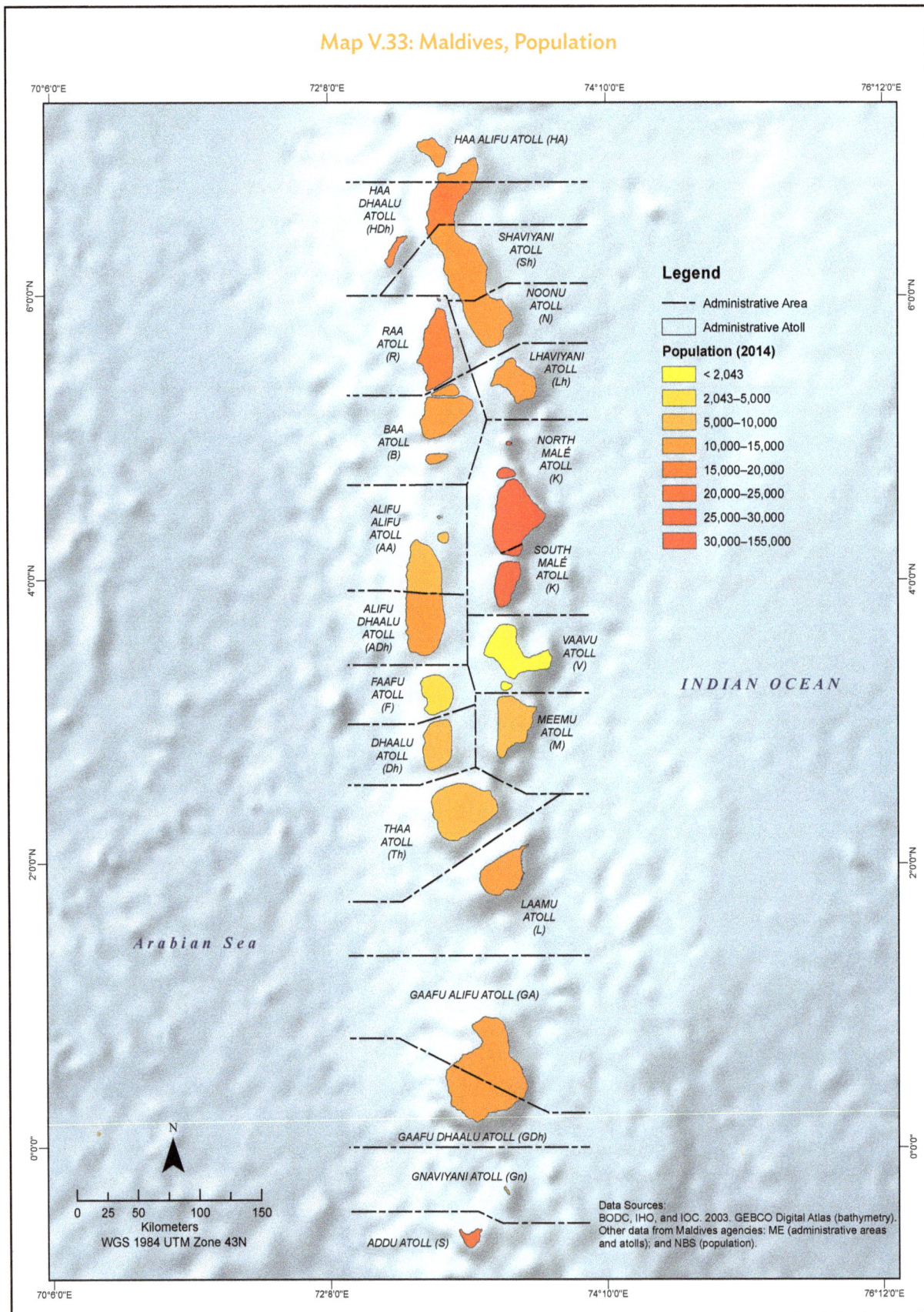

Legend

– – – Administrative Area

▭ Administrative Atoll

Population (2014)

🟨	< 2,043
🟨	2,043–5,000
🟧	5,000–10,000
🟧	10,000–15,000
🟧	15,000–20,000
🟥	20,000–25,000
🟥	25,000–30,000
🟥	30,000–155,000

HAA ALIFU ATOLL (HA)

HAA DHAALU ATOLL (HDh)

SHAVIYANI ATOLL (Sh)

NOONU ATOLL (N)

RAA ATOLL (R)

LHAVIYANI ATOLL (Lh)

BAA ATOLL (B)

NORTH MALÉ ATOLL (K)

ALIFU ALIFU ATOLL (AA)

SOUTH MALÉ ATOLL (K)

ALIFU DHAALU ATOLL (ADh)

VAAVU ATOLL (V)

FAAFU ATOLL (F)

MEEMU ATOLL (M)

DHAALU ATOLL (Dh)

THAA ATOLL (Th)

LAAMU ATOLL (L)

GAAFU ALIFU ATOLL (GA)

GAAFU DHAALU ATOLL (GDh)

GNAVIYANI ATOLL (Gn)

ADDU ATOLL (S)

INDIAN OCEAN

Arabian Sea

N

0 25 50   100   150
Kilometers
WGS 1984 UTM Zone 43N

Data Sources:
BODC, IHO, and IOC. 2003. GEBCO Digital Atlas (bathymetry).
Other data from Maldives agencies: ME (administrative areas
and atolls); and NBS (population).

Map V.34: Maldives, Population Density

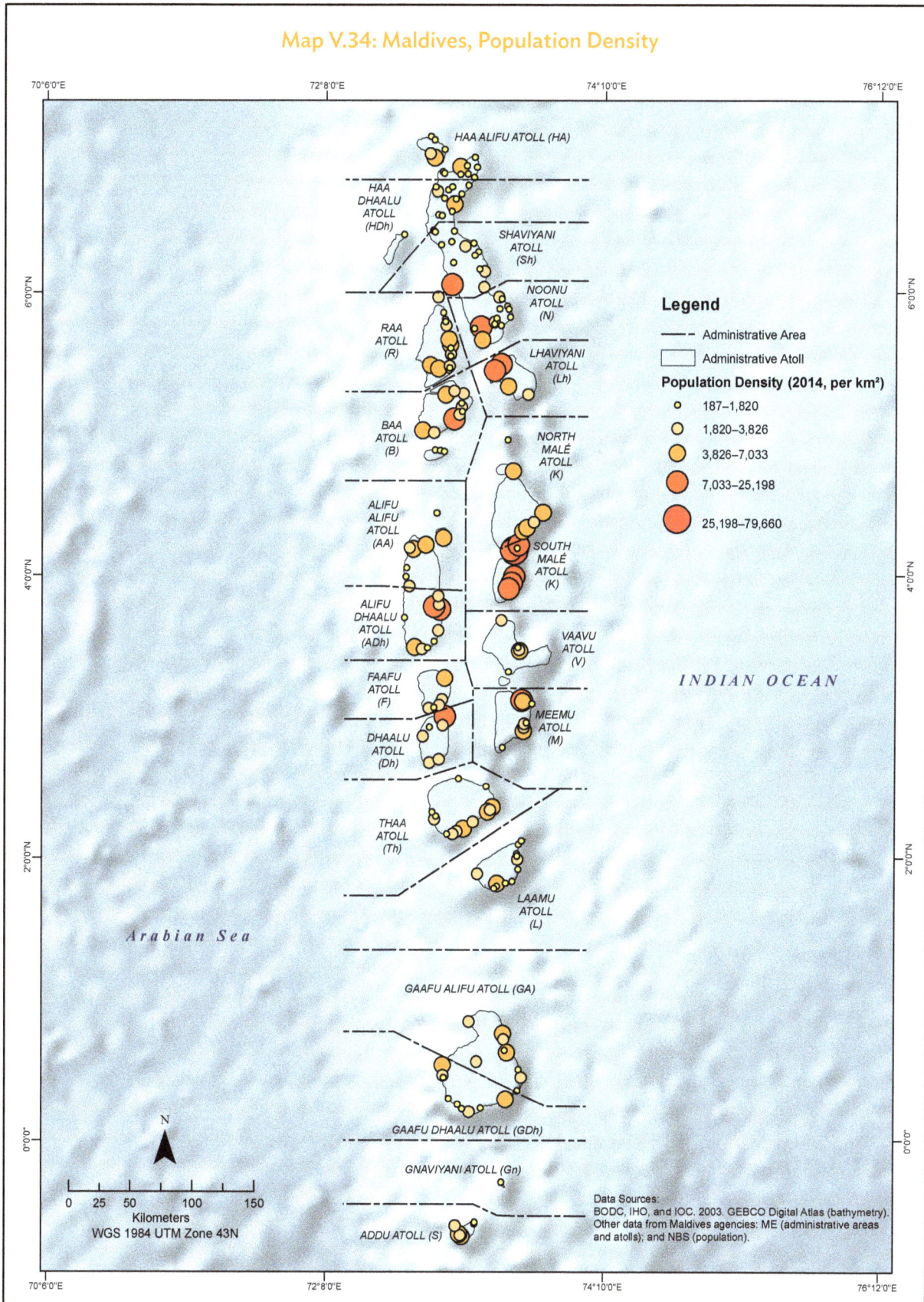

HAA ALIFU ATOLL (HA)

HAA DHAALU ATOLL (HDh)

SHAVIYANI ATOLL (Sh)

NOONU ATOLL (N)

RAA ATOLL (R)

LHAVIYANI ATOLL (Lh)

BAA ATOLL (B)

NORTH MALÉ ATOLL (K)

ALIFU ALIFU ATOLL (AA)

SOUTH MALÉ ATOLL (K)

ALIFU DHAALU ATOLL (ADh)

VAAVU ATOLL (V)

FAAFU ATOLL (F)

MEEMU ATOLL (M)

DHAALU ATOLL (Dh)

THAA ATOLL (Th)

LAAMU ATOLL (L)

GAAFU ALIFU ATOLL (GA)

GAAFU DHAALU ATOLL (GDh)

GNAVIYANI ATOLL (Gn)

ADDU ATOLL (S)

INDIAN OCEAN

Arabian Sea

Legend

– – – Administrative Area

☐ Administrative Atoll

Population Density (2014, per km²)

- 187–1,820
- 1,820–3,826
- 3,826–7,033
- 7,033–25,198
- 25,198–79,660

N

0 25 50 100 150
Kilometers
WGS 1984 UTM Zone 43N

Data Sources:
BODC, IHO, and IOC. 2003. GEBCO Digital Atlas (bathymetry).
Other data from Maldives agencies: ME (administrative areas
and atolls); and NBS (population).

Education

Maldives has a high literacy rate. About 88,000 students are enrolled in the country's 376 government, private, and community schools, most of which are in Kaafu Atoll. Less than 2% of the country's population are illiterate and less than 1% of the youth aged 15–24 cannot read and write (United Nations Educational, Scientific and Cultural Organization's Institute for Statistics 2014).

Schools with the highest student capacities are located in Malé, which is also in Kaafu Atoll and the city with the largest population in Maldives. Most of the schools in Malé are situated in low-lying grounds where they are vulnerable to various hazards related to inundation.

Religion and literacy. All Maldivians practice Islam as a religion as required by law. Almost all people in Maldives are literate (right photo by Mark Fischer, left photo by Adam Jones).

Table V.7: Number of Schools and Student Capacity per Atoll, Maldives

Atoll	Number of Schools	Student Capacity
Alifu Alifu	13	1,681
Alifu Dhaalu	18	2,130
Baa	22	2,614
Dhaalu	11	1,697
Faafu	8	1,421
Gaafu Alifu	14	2,242
Gaafu Dhaalu	17	3,197
Gnaviyani	8	2,465
Haa Alifu	24	3,584
Haa Dhaalu	29	5,625
Kaafu	54	34,440
Laamu	21	3,527
Lhaviyani	7	2,092
Meemu	12	1,217
Noonu	25	2,967
Raa	28	4,704
Seenu	22	5,611
Shaviyani	19	3,680
Thaa	19	2,547
Vaavu	5	347
TOTAL	**376**	**87,788**

Source: Ministry of Education, 2016.

Health

Changing climate—particularly extreme weather conditions—and geologic hazards can affect the health of people, especially persons with disabilities, the elderly, children, and pregnant women. These can cause injuries and even death. Warmer temperature and wetter monsoons, as seen in climate projections for Maldives *(Multihazard Risk Atlas of Maldives: Climate and Geophysical Hazards—Volume II)*, could increase the incidences of vector-borne diseases like dengue (Ahmed and Suphachalasai 2014) and water-borne diseases (Intergovernmental Panel on Climate Change 2001). Inundation, as a result of elevated sea levels or rainfall-induced flooding, could generate an environment favorable to breeding of mosquitoes and other disease vectors (Ahmed and Suphachalasai 2014).

This section maps out health facilities as one indicator of vulnerability to climate and disaster risks. Based on 2016 data from the Ministry of Health, Maldives has a total of 184 hospitals classified into health centers, private hospitals, government hospitals, atoll hospitals, and regional hospitals. These hospitals are distributed mostly among the highly populated atolls of Raa, Kaafu, and Haa Alifu.

Table V.8: Types of Hospitals in Maldives

Hospital Type	Number
Health center	161
Atoll hospital	13
Regional hospital	6
Government hospital	2
Private hospital	2

Source: Ministry of Health, 2016.

Table V.9: Number of Hospitals per Atoll, Maldives

ATOLL	Number of Hospitals
Alifu Dhaalu Atoll	1
Alifu Alifu Atoll	8
Alifu Dhaalu Atoll	9
Baa Atoll	13
Dhaalu Atoll	7
Faafu Atoll	5
Gaafu Alifu Atoll	8
Gaafu Dhaalu Atoll	9
Gnaviyani Atoll	1
Haa Alifu Atoll	14
Haa Dhaalu Atoll	13
Kaafu Atoll	14
Laamu Atoll	11
Lhaviyani Atoll	4
Meemu Atoll	8
Noonu Atoll	12
Raa Atoll	15
Seenu Atoll	4
Shaviyani Atoll	14
Thaa Atoll	13
Vaavu Atoll	1
TOTAL	**184**

Source: Ministry of Health, 2016.

Map V.35: Maldives, Healthcare Facilities

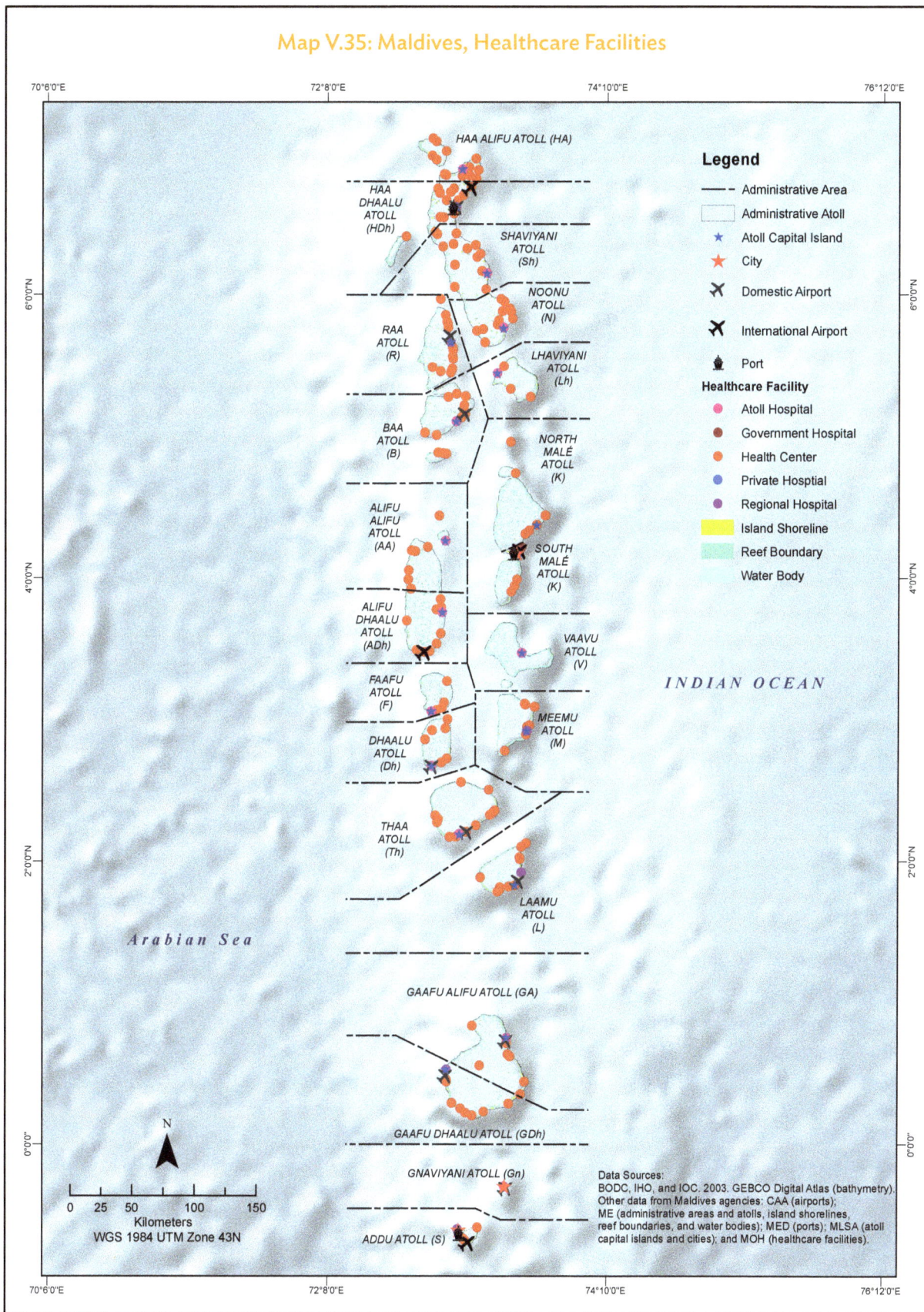

Legend

- – – – Administrative Area
- ☐ Administrative Atoll
- ★ Atoll Capital Island
- ★ City
- ✈ Domestic Airport
- ✈ International Airport
- ⚓ Port

Healthcare Facility
- ● Atoll Hospital
- ● Government Hospital
- ● Health Center
- ● Private Hosptial
- ● Regional Hospital
- Island Shoreline
- Reef Boundary
- Water Body

HAA ALIFU ATOLL (HA)
HAA DHAALU ATOLL (HDh)
SHAVIYANI ATOLL (Sh)
NOONU ATOLL (N)
RAA ATOLL (R)
LHAVIYANI ATOLL (Lh)
BAA ATOLL (B)
NORTH MALÉ ATOLL (K)
ALIFU ALIFU ATOLL (AA)
SOUTH MALÉ ATOLL (K)
ALIFU DHAALU ATOLL (ADh)
VAAVU ATOLL (V)
FAAFU ATOLL (F)
MEEMU ATOLL (M)
DHAALU ATOLL (Dh)
THAA ATOLL (Th)
LAAMU ATOLL (L)
GAAFU ALIFU ATOLL (GA)
GAAFU DHAALU ATOLL (GDh)
GNAVIYANI ATOLL (Gn)
ADDU ATOLL (S)

INDIAN OCEAN

Arabian Sea

N

0 25 50 100 150
Kilometers
WGS 1984 UTM Zone 43N

Data Sources:
BODC, IHO, and IOC. 2003. GEBCO Digital Atlas (bathymetry).
Other data from Maldives agencies: CAA (airports);
ME (administrative areas and atolls, island shorelines,
reef boundaries, and water bodies); MED (ports); MLSA (atoll
capital islands and cities); and MOH (healthcare facilities).

Lives Centered at Sea

Life in Maldives revolves around the sea, which provides Maldivians with plenty of resources for livelihood including fisheries and tourism.

Rich aquatic resources made fisheries the main source of livelihood in Maldives, until tourism took over. Up to now, fishing contributes to national productivity with tuna, grouper, and snapper among the popular species of fish caught and sold in the market as food.

The natural beauty of its beaches has enticed tourists to visit Maldives for over 4 decades. Although the sea and tropical climate provide for the success of the tourism sector, they also pose concerns for changes in the country's sea level and future climate.

Fisheries and tourism are susceptible to the changing climate. Future warmer temperatures can affect fish production as higher sea surface temperature leads to coral bleaching, and tourists wanting to see the rich aquatic life in Maldives may no longer be as satisfied (Hosterman and Smith 2014). A wetter climate can also affect the influx of tourists.

In the future, a wetter climate could be less favorable for tourism activities (*Multihazard Risk Atlas of Maldives: Climate and Geophysical Hazards—Volume II*, p. 32). This may be observed especially during the southwest monsoon in June, July, and August in northern Maldives (*Multihazard Risk Atlas of Maldives: Climate and Geophysical Hazards—Volume II*, p. 47). The beaches are also threatened by coastal erosion (*Multihazard Risk Atlas of Maldives: Biodiversity—Volume IV*, p. 69). Additionally, low-lying islands are prone to inundation.

An atoll as seen from the sky. A portion of a Maldivian island is captured on an aerial photograph (photo by Sue Todd).

Tourism

Naturally beautiful beaches and the presence of island resorts make tourism the main economic sector in Maldives. Tourists can choose to visit one of the 120 islands with operating resorts, ranging from small resorts with only 14 beds to larger ones with more than 900 beds. Most resorts are located in and around Kaafu Atoll (43 resorts), Alifu Dhaalu Atoll (17 resorts), and Baa Atoll (12 resorts) due to the proximity to Malé City and international airports.

Baa, Lhaviyani, Haa Alifu, and Shaviyani atolls will experience an increase in average rainfall rate, based on climate projection results (*Multihazard Risk Atlas of Maldives: Climate and Geophysical Hazards—Volume II*, p. 32). The resorts in these atolls may experience higher rainfall rate in the future.

Table V.10: Number of Resorts per Atoll, Maldives

Atoll	Number of Resorts
Kaafu	43
Alifu Dhaalu	17
Baa	12
Alifu Alifu	11
Lhaviyani	7
Dhaalu	5
Noonu	5
Gaafu Alifu	4
Gaafu Dhaalu	3
Raa	3
Addu City	2
Meemu	2
Vaavu	2
Faafu	1
Haa Alifu	1
Laamu	1
Thaa	1

Source: Ministry of Tourism, 2017.

Map V.36: Maldives, Resort Islands

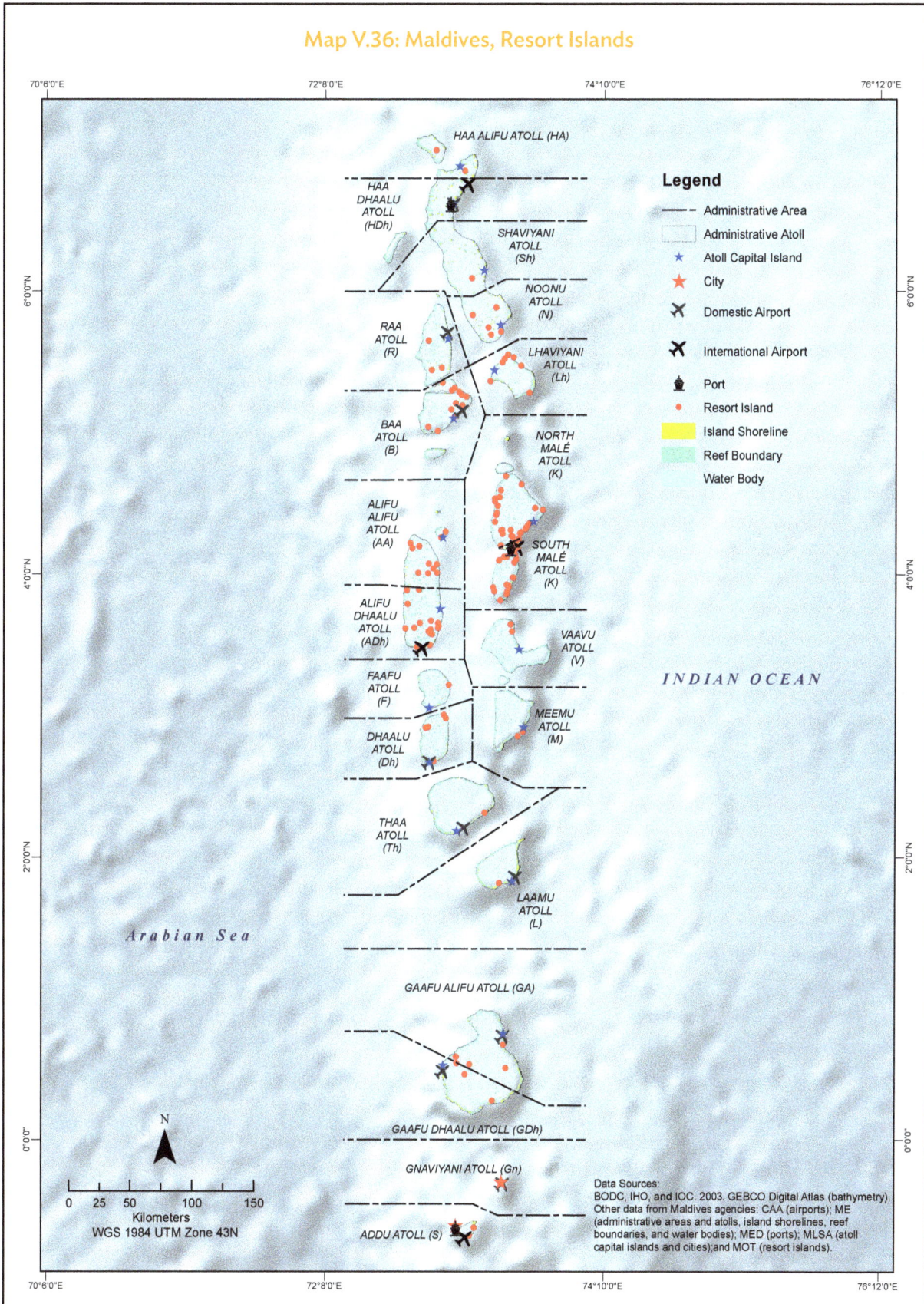

70°6'0"E   72°8'0"E   74°10'0"E   76°12'0"E

HAA ALIFU ATOLL (HA)

HAA DHAALU ATOLL (HDh)

SHAVIYANI ATOLL (Sh)

NOONU ATOLL (N)

RAA ATOLL (R)

LHAVIYANI ATOLL (Lh)

BAA ATOLL (B)

NORTH MALÉ ATOLL (K)

ALIFU ALIFU ATOLL (AA)

SOUTH MALÉ ATOLL (K)

ALIFU DHAALU ATOLL (ADh)

VAAVU ATOLL (V)

FAAFU ATOLL (F)

MEEMU ATOLL (M)

DHAALU ATOLL (Dh)

INDIAN OCEAN

THAA ATOLL (Th)

LAAMU ATOLL (L)

Arabian Sea

GAAFU ALIFU ATOLL (GA)

GAAFU DHAALU ATOLL (GDh)

GNAVIYANI ATOLL (Gn)

ADDU ATOLL (S)

Legend

- — — Administrative Area
- ⬚ Administrative Atoll
- ★ Atoll Capital Island
- ★ City
- ✕ Domestic Airport
- ✈ International Airport
- ⚓ Port
- ● Resort Island
- ▬ Island Shoreline
- ▬ Reef Boundary
- Water Body

N

0 25 50 100 150
Kilometers
WGS 1984 UTM Zone 43N

Data Sources:
BODC, IHO, and IOC. 2003. GEBCO Digital Atlas (bathymetry).
Other data from Maldives agencies: CAA (airports); ME
(administrative areas and atolls, island shorelines, reef
boundaries, and water bodies); MED (ports); MLSA (atoll
capital islands and cities);and MOT (resort islands).

N°0.0°9   N°0.0°9

N°0.0°4   N°0.0°4

N°0.0°2   N°0.0°2

0°0.0°   0°0.0°

70°6'0"E   72°8'0"E   74°10'0"E   76°12'0"E

Transportation

In Maldives, traveling from one end of the island to the other does not take too long, and there are various modes of transportation. People can choose to travel by air or sea.

There are three ports for traveling and transporting goods: (i) Kulhudhuffushi Regional Port in Haa Dhaalu, (ii) Hithadhoo Regional Port in Addu City, and (iii) Malé Commercial Harbor in Malé City. People can also board planes through the international and domestic airports. One alternative mode of transport is via floatplane, which can be accessed from floating platforms scattered across the islands.

Means of transport. A seaplane docked at a port in Maldives is a usual sight and transport mode (photo by Ibahim Asad).

Table V.11: Airports in Maldives

Atoll	Island Name	Aerodrome
Domestic Airports		
Baa	Dharavandhoo	Dharavandhoo Airport
Gnaviyani	Fuvahmulah	Fuvahmulah Airport
Dhaalu	Kudahuvadhoo	Dhaalu Airport
Thaa	Thimarafushi	Thimarafushi Airport
Laamu	Kahdhoo	Kadhdhoo Airport
Gaafu Dhaalu	Kaadehdhoo	Kaadedhdhoo Airport
Alifu Dhaalu	Maamin'gili	Villa Airport Maamigili
Gaafu Alifu	Kooddoo	Koodoo Airport
Raa	Ifuru	Ifuru Airport
Haa Dhaalu	Kulhudhuffushi	Kulhudhuffushi Airport
International Airports		
Noonu	Maafaru	Maafaru International Airport
Haa Dhaalu	Hanimaadhoo	Hanimaadhoo International Airport
Kaafu	Hulhulé	Velana International Airport
Seenu	Gan	Gan International Airport
Future Airports		
Gaafu Dhaalu	Faresmaathodaa	
Gaafu Dhaalu	Maavaarulu	
Haa Alifu	Hoarafushi	
Lhaviyani	Madivaru	
Meemu	Muli	
Shaviyani	Funadhoo	

Source: Civil Aviation Authority, 2019.

Map V.37: Maldives, Transportation

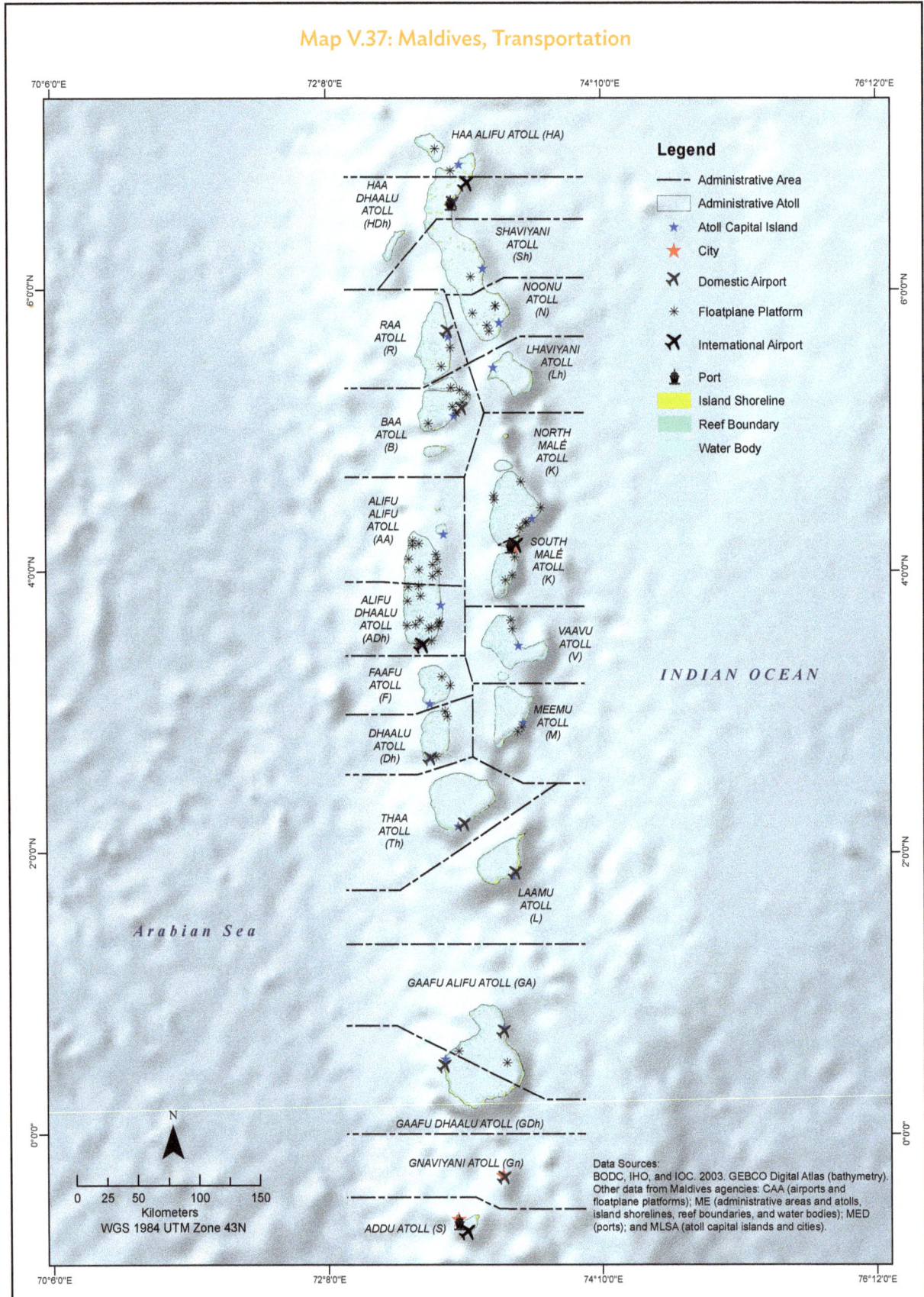

Legend

- — — Administrative Area
- ☐ Administrative Atoll
- ★ Atoll Capital Island
- ★ City
- ✕ Domestic Airport
- ✳ Floatplane Platform
- ✕ International Airport
- ⚓ Port
- ▮ Island Shoreline
- ▮ Reef Boundary
- ▮ Water Body

HAA ALIFU ATOLL (HA)

HAA DHAALU ATOLL (HDh)

SHAVIYANI ATOLL (Sh)

NOONU ATOLL (N)

RAA ATOLL (R)

LHAVIYANI ATOLL (Lh)

BAA ATOLL (B)

NORTH MALÉ ATOLL (K)

ALIFU ALIFU ATOLL (AA)

SOUTH MALÉ ATOLL (K)

ALIFU DHAALU ATOLL (ADh)

VAAVU ATOLL (V)

FAAFU ATOLL (F)

MEEMU ATOLL (M)

DHAALU ATOLL (Dh)

INDIAN OCEAN

THAA ATOLL (Th)

LAAMU ATOLL (L)

Arabian Sea

GAAFU ALIFU ATOLL (GA)

N

GAAFU DHAALU ATOLL (GDh)

GNAVIYANI ATOLL (Gn)

ADDU ATOLL (S)

0 25 50   100   150
Kilometers
WGS 1984 UTM Zone 43N

Data Sources:
BODC, IHO, and IOC. 2003. GEBCO Digital Atlas (bathymetry).
Other data from Maldives agencies: CAA (airports and
floatplane platforms); ME (administrative areas and atolls,
island shorelines, reef boundaries, and water bodies); MED
(ports); and MLSA (atoll capital islands and cities).

70°6'0"E   72°8'0"E   74°10'0"E   76°12'0"E

6°0'0"N   4°0'0"N   2°0'0"N   0°0'0"

Harbor Facilities

For an archipelagic country like Maldives, harbors are important nodes in the transportation system. People can mobilize through these harbors. However, not all the inhabited islands have harbor facilities.

Boats known locally as *dhoni* and ferries are other modes of transportation in Maldives. They dock in harbor facilities installed along the coasts of 170 islands, 17 of which are atoll capitals. Out of 199 harbors in the country, Raa and Shaviyani atolls have the most with 15 harbors each.

Traditional boats. Traditional boats called *dhoni* docked at a port in Maldives (photo by Azwar Thaufeeq).

Table V.12: Harbors in Maldives

Atoll	Number of Harbors
Raa Atoll	15
Shaviyani Atoll	15
Laamu Atoll	14
Haa Dhaalu Atoll	14
Haa Alifu Atoll	13
Baa Atoll	13
Malé City	13
Thaa Atoll	11
Gaafu Alifu Atoll	10
Alifu Dhaalu Atoll	10
Noonu Atoll	10
Gaafu Dhaalu Atoll	9
Alifu Alifu Atoll	8
Kaafu Atoll	8
Meemu Atoll	7
Dhaalu Atoll	6
Faafu Atoll	6
Lhaviyani Atoll	6
Addu City	6
Vaavu Atoll	4
Gnaviyani Atoll	1

Source: Ministry of National Planning and Infrastructure, 2017.

Common port scene. Small motorboats are a usual sight at ports in Maldives (photo by Department of Foreign Affairs and Trade).

Map V.38: Maldives, Harbor Facilities

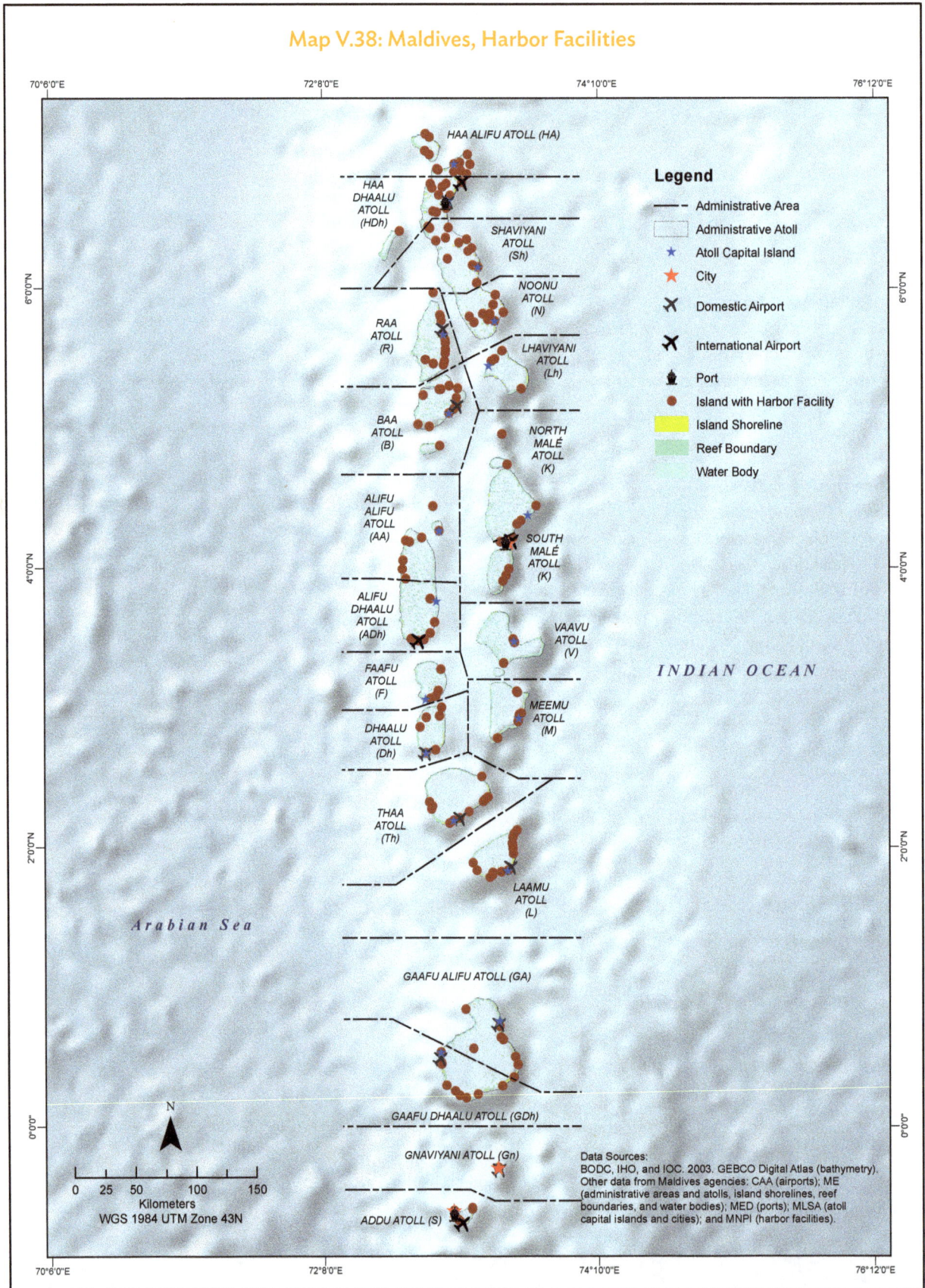

Legend

- – – Administrative Area
- ☐ Administrative Atoll
- ★ Atoll Capital Island
- ★ City
- ✈ Domestic Airport
- ✈ International Airport
- ⚓ Port
- ● Island with Harbor Facility
- Island Shoreline
- Reef Boundary
- Water Body

HAA ALIFU ATOLL (HA)

HAA DHAALU ATOLL (HDh)

SHAVIYANI ATOLL (Sh)

NOONU ATOLL (N)

RAA ATOLL (R)

LHAVIYANI ATOLL (Lh)

BAA ATOLL (B)

NORTH MALÉ ATOLL (K)

ALIFU ALIFU ATOLL (AA)

SOUTH MALÉ ATOLL (K)

ALIFU DHAALU ATOLL (ADh)

VAAVU ATOLL (V)

FAAFU ATOLL (F)

MEEMU ATOLL (M)

DHAALU ATOLL (Dh)

THAA ATOLL (Th)

LAAMU ATOLL (L)

GAAFU ALIFU ATOLL (GA)

GAAFU DHAALU ATOLL (GDh)

GNAVIYANI ATOLL (Gn)

ADDU ATOLL (S)

INDIAN OCEAN

Arabian Sea

N

Kilometers
0 25 50   100   150
WGS 1984 UTM Zone 43N

Data Sources:
BODC, IHO, and IOC. 2003. GEBCO Digital Atlas (bathymetry).
Other data from Maldives agencies: CAA (airports); ME
(administrative areas and atolls, island shorelines, reef
boundaries, and water bodies); MED (ports); MLSA (atoll
capital islands and cities); and MNPI (harbor facilities).

Sand Mining Applications

Sand is one of the major natural resources in Maldives. The white sand lining shores is attributed to the coralline formation of the islands.

In the country, sand mining is a common practice—mined sand is used for construction purposes as there is no other source of aggregate material. In recent years, coralline sand mining has increased to meet the demand of development (Naseer 1997). However, the high demand threatens beaches and islands. Sand mining activities could damage the integrity of shores and lead to erosion (Emerton, Baig, and Saleem 2009). It also threatens corals and other aquatic organisms (Ministry of Environment and Energy 2016). Additionally, the extraction of sand alters the natural movement of sediments and environmental processes on which organisms rely (Dhunya, Huang, and Aslam 2017).

To protect corals, sand mining is regulated through the Act on Sand Mining of 1978, which requires sand mining permits, and the Act on Coral and Sand Mining of 2000, which regulates sand mining only in designated locations for those who have permits.

There are currently 157 approved sand mining locations in Maldives, the majority of which are in Laamu (15), Baa (13), and Thaa (13) atolls (Table V.13).

Table V.13: Sand Mining in Maldives

Atoll	Number of Sand Mining Locations
Laamu	15
Baa	13
Thaa	13
Gaafu Dhaalu	12
Kaafu	11
Raa	9
Haa Alifu	9
Gaafu Alifu	9
Alifu Dhaalu	8
Dhaalu	8
Meemu	8
Haa Dhaalu	6
Alifu Alifu	6
Seenu	6
Laviyani	5
Shaviyani	5
Noonu	5
Faffu	5
Vaavu	4

Source: International Union for Conservation of Nature, Maldives, 2016.

Mineral source. Sand are mined in Maldives for construction purposes (photo by Sue Todd).

Map V.39: Maldives, Approved Sand Mining Locations

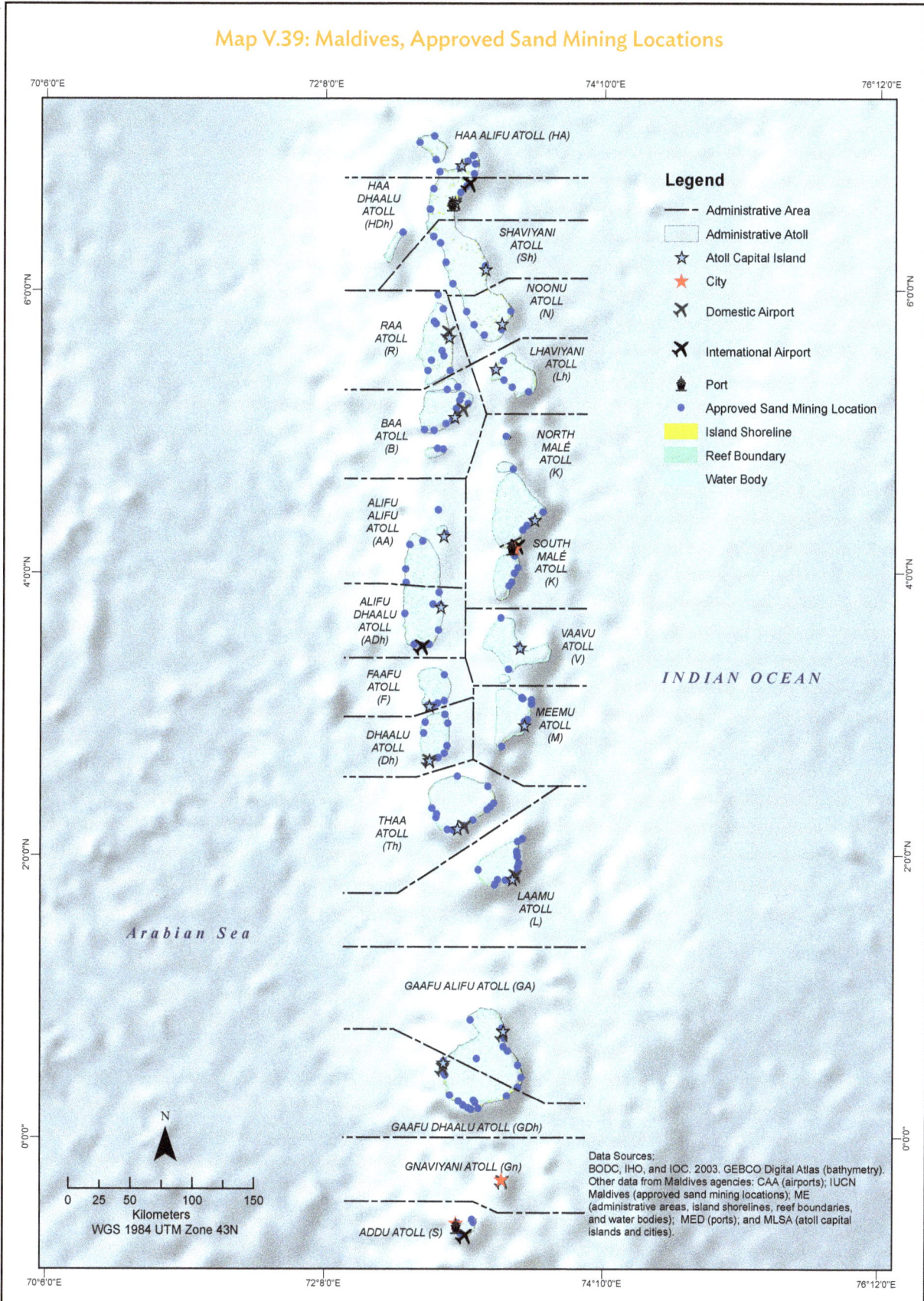

HAA ALIFU ATOLL (HA)

HAA DHAALU ATOLL (HDh)

SHAVIYANI ATOLL (Sh)

NOONU ATOLL (N)

RAA ATOLL (R)

LHAVIYANI ATOLL (Lh)

BAA ATOLL (B)

NORTH MALÉ ATOLL (K)

ALIFU ALIFU ATOLL (AA)

SOUTH MALÉ ATOLL (K)

ALIFU DHAALU ATOLL (ADh)

VAAVU ATOLL (V)

FAAFU ATOLL (F)

MEEMU ATOLL (M)

DHAALU ATOLL (Dh)

THAA ATOLL (Th)

LAAMU ATOLL (L)

GAAFU ALIFU ATOLL (GA)

GAAFU DHAALU ATOLL (GDh)

GNAVIYANI ATOLL (Gn)

ADDU ATOLL (S)

INDIAN OCEAN

Arabian Sea

Legend

- Administrative Area
- Administrative Atoll
- ☆ Atoll Capital Island
- ★ City
- Domestic Airport
- International Airport
- Port
- • Approved Sand Mining Location
- Island Shoreline
- Reef Boundary
- Water Body

N

0 25 50 100 150
Kilometers
WGS 1984 UTM Zone 43N

Data Sources:
BODC, IHO, and IOC. 2003. GEBCO Digital Atlas (bathymetry).
Other data from Maldives agencies: CAA (airports); IUCN
Maldives (approved sand mining locations); ME
(administrative areas, island shorelines, reef boundaries,
and water bodies); MED (ports); and MLSA (atoll capital
islands and cities).

70°6'0"E   72°8'0"E   74°10'0"E   76°12'0"E

6°0'0"N

4°0'0"N

2°0'0"N

0°0'0"

Power Stations

All the inhabited islands of Maldives have access to continuous electric supply. Two major electric supply companies—Fenaka Corporation Limited and State Electric Company Limited—provide electricity to the islands. The largest electric demand comes from Malé City.

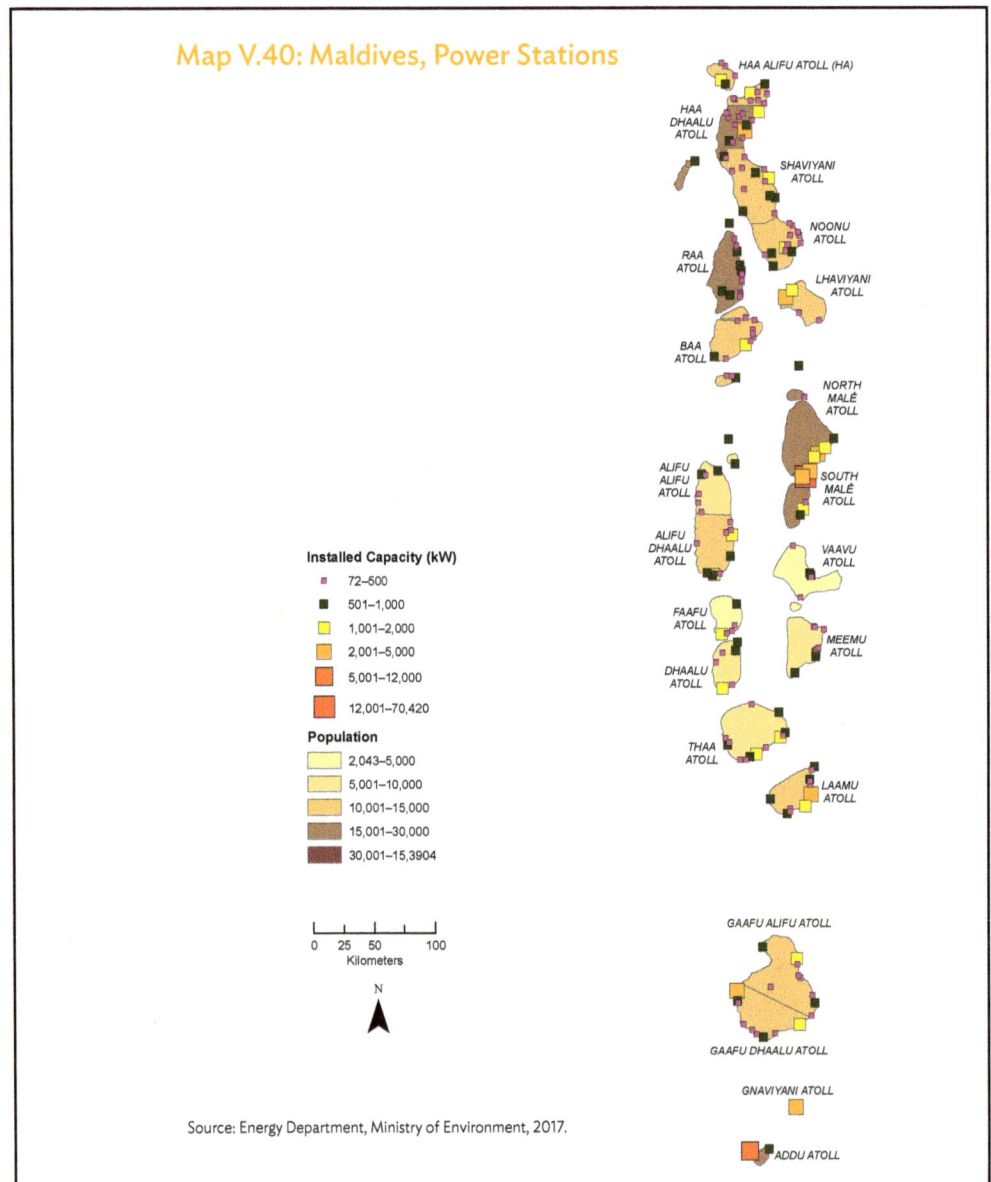

Map V.40: Maldives, Power Stations

Installed Capacity (kW)
- 72–500
- 501–1,000
- 1,001–2,000
- 2,001–5,000
- 5,001–12,000
- 12,001–70,420

Population
- 2,043–5,000
- 5,001–10,000
- 10,001–15,000
- 15,001–30,000
- 30,001–15,3904

Source: Energy Department, Ministry of Environment, 2017.

Human Development Index

The Human Development Index is a summary measure developed by the United Nations Development Programme to assess national development based on three basic dimensions of human development: a long and healthy life, access to knowledge, and a decent standard of living (UNDP 2018). In Maldives, ratings are high for Faafu, Dhaalu, and Meemu atolls as well as Malé City.

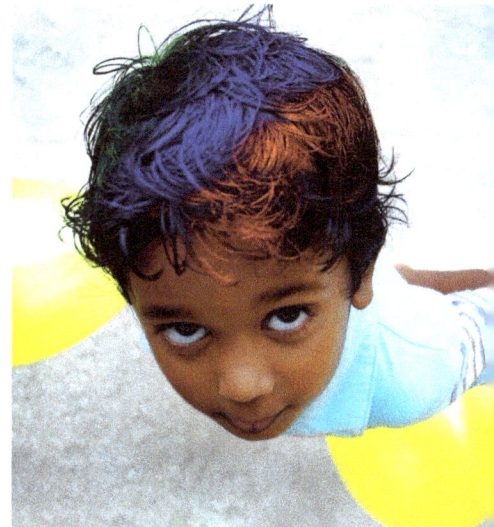

Maldivians on a sunny day at the beach. Human development include activities and interactions with other people and relationship with the environment (photos by Sue Todd).

Map V.41: Maldives, Human Development Index

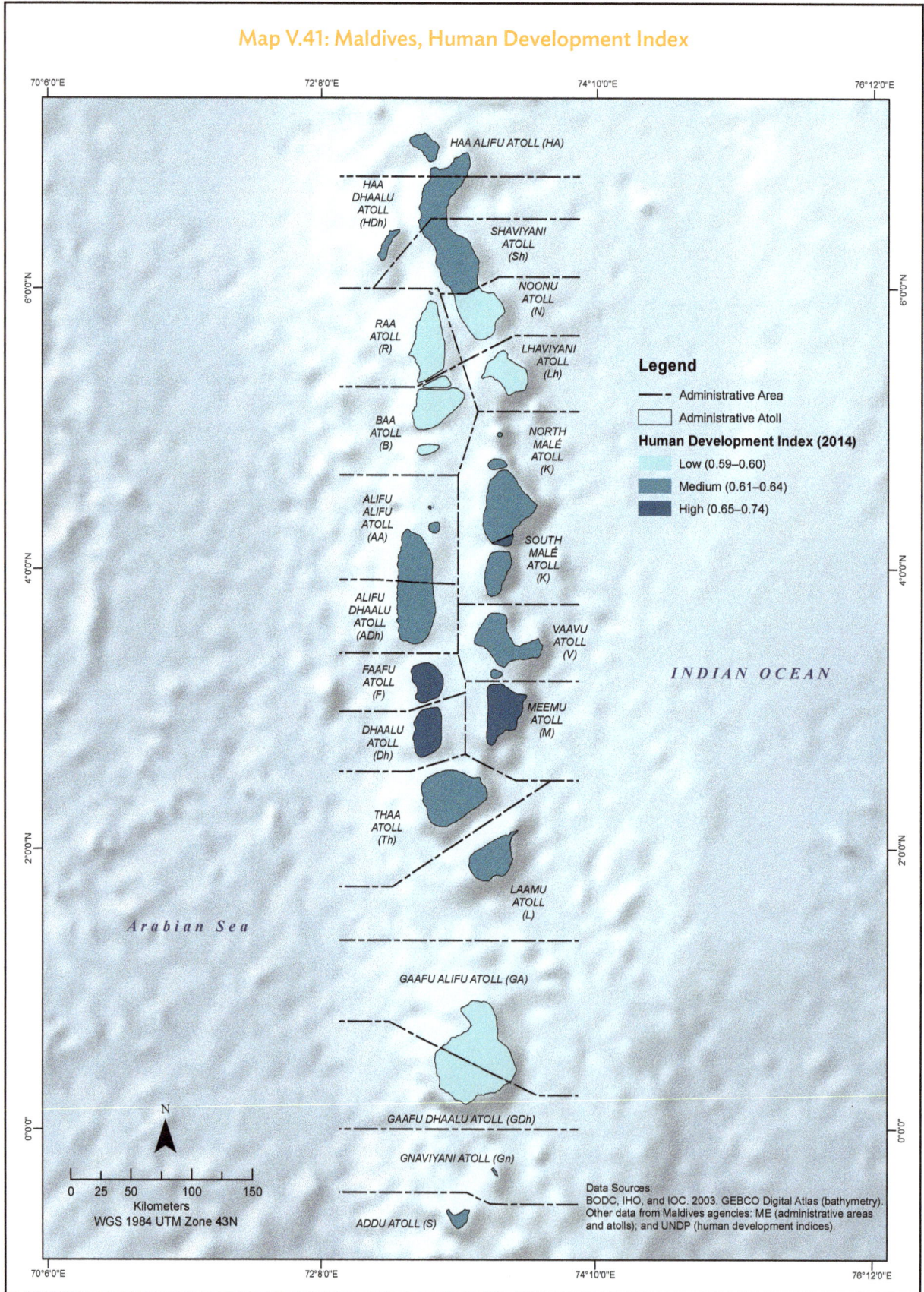

HAA ALIFU ATOLL (HA)

HAA DHAALU ATOLL (HDh)

SHAVIYANI ATOLL (Sh)

NOONU ATOLL (N)

RAA ATOLL (R)

LHAVIYANI ATOLL (Lh)

BAA ATOLL (B)

NORTH MALÉ ATOLL (K)

ALIFU ALIFU ATOLL (AA)

SOUTH MALÉ ATOLL (K)

ALIFU DHAALU ATOLL (ADh)

VAAVU ATOLL (V)

FAAFU ATOLL (F)

MEEMU ATOLL (M)

DHAALU ATOLL (Dh)

THAA ATOLL (Th)

LAAMU ATOLL (L)

INDIAN OCEAN

Arabian Sea

GAAFU ALIFU ATOLL (GA)

GAAFU DHAALU ATOLL (GDh)

GNAVIYANI ATOLL (Gn)

ADDU ATOLL (S)

Legend

– – – Administrative Area

☐ Administrative Atoll

Human Development Index (2014)

Low (0.59–0.60)

Medium (0.61–0.64)

High (0.65–0.74)

N

0 25 50 100 150
Kilometers
WGS 1984 UTM Zone 43N

Data Sources:
BODC, IHO, and IOC. 2003. GEBCO Digital Atlas (bathymetry).
Other data from Maldives agencies: ME (administrative areas
and atolls); and UNDP (human development indices).

Environmentally Sensitive Areas

Maldives is naturally blessed with an environment that supports a variety of terrestrial and aquatic life. As of 2017, it has a total of 284 environmentally sensitive areas. A quarter of these sites are for bird assembly, nesting, and habitat. Migratory birds find temporary refuge in the islands of Shaviyani Atoll. The mangrove ecosystem, which exists mostly in islands with the least human settlements, serves as breeding grounds for sharks and rays and as a nesting site for turtles. In addition to providing protection to the coast and coastal communities, mangroves also protect the white-breasted waterhen (kabili) and tortoise and serve as shelters for aquatic organisms.

Having evolved from corals, Maldives also has black, soft, hard, and table corals that shelter a variety of marine life. Divers visit Maldives to experience these beautiful coral formations and marine ecosystems teeming with life such as the yellowback fusilier, humpback snapper, whitetip shark, whale shark, grey shark, nurse shark, leopard shark, guitar shark, hammerhead shark, needlefish, eagle ray, manta ray, turtle, tortoise, sea cucumber, lionfish, moray eel, tuna, barracuda, sailfish, red snapper, grouper, and many others. Maldives also has local medicinal plants that the country aims to protect.

Rich biodiversity in Maldives. (Clockwise from top right) The *kabili*, also known as the white-breasted waterhen, is the national bird, while the grey heron is a common sight in some coasts (photos by Wang Chaonan and Anastasia Kolchedantseva). Under the sea, blacktip reef sharks and hawksbill sea turtles abound (photos by Ibrahim Rifath and Andrew Corman).

Map V.42: Maldives, Environmentally Protected and Sensitive Areas

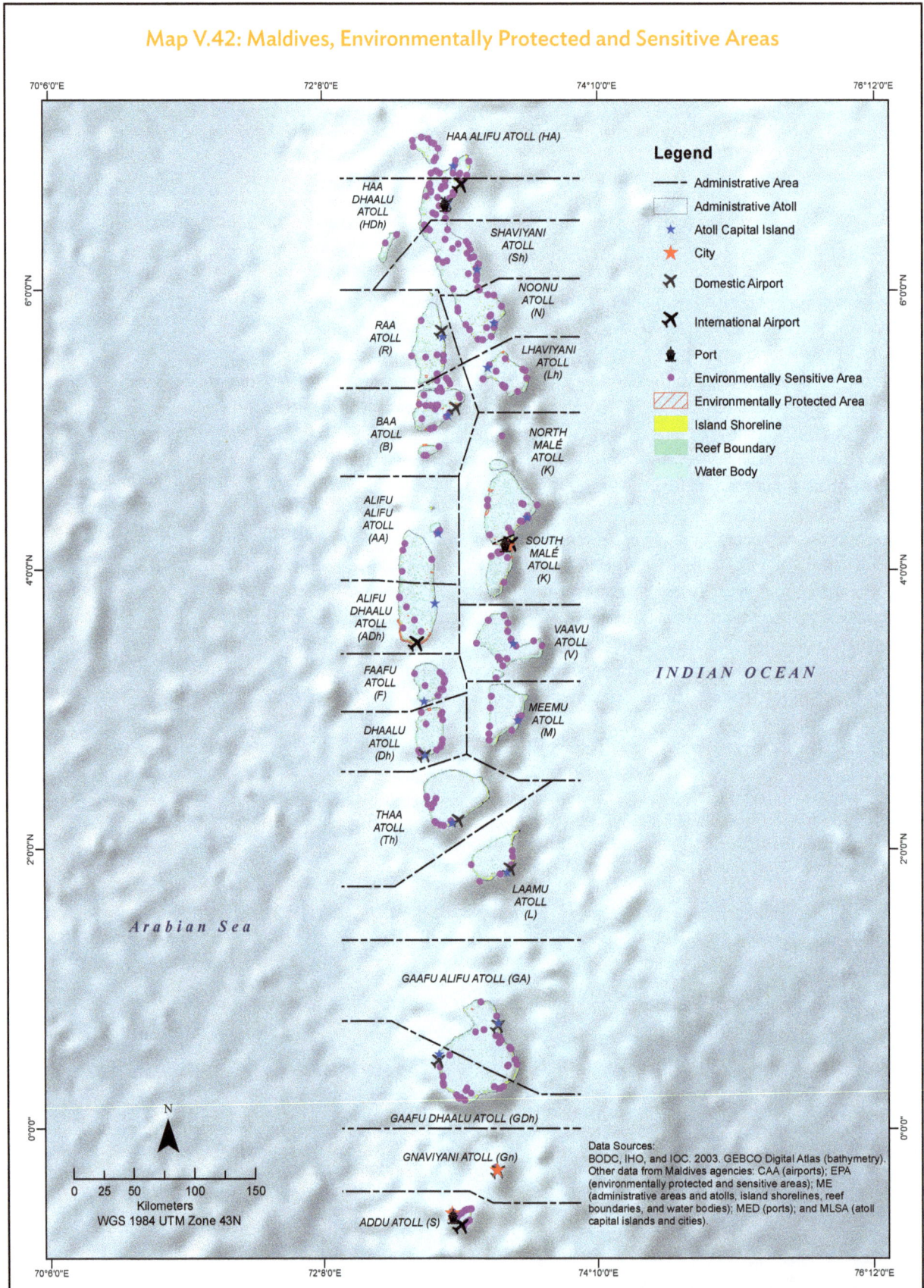

Legend

- Administrative Area
- Administrative Atoll
- ★ Atoll Capital Island
- ★ City
- ✈ Domestic Airport
- ✈ International Airport
- ⚓ Port
- ● Environmentally Sensitive Area
- Environmentally Protected Area
- Island Shoreline
- Reef Boundary
- Water Body

HAA ALIFU ATOLL (HA)

HAA DHAALU ATOLL (HDh)

SHAVIYANI ATOLL (Sh)

NOONU ATOLL (N)

RAA ATOLL (R)

LHAVIYANI ATOLL (Lh)

BAA ATOLL (B)

NORTH MALÉ ATOLL (K)

ALIFU ALIFU ATOLL (AA)

SOUTH MALÉ ATOLL (K)

ALIFU DHAALU ATOLL (ADh)

VAAVU ATOLL (V)

FAAFU ATOLL (F)

MEEMU ATOLL (M)

DHAALU ATOLL (Dh)

THAA ATOLL (Th)

LAAMU ATOLL (L)

GAAFU ALIFU ATOLL (GA)

GAAFU DHAALU ATOLL (GDh)

GNAVIYANI ATOLL (Gn)

ADDU ATOLL (S)

INDIAN OCEAN

Arabian Sea

Kilometers 0 25 50 100 150
WGS 1984 UTM Zone 43N

Data Sources:
BODC, IHO, and IOC. 2003. GEBCO Digital Atlas (bathymetry). Other data from Maldives agencies: CAA (airports); EPA (environmentally protected and sensitive areas); ME (administrative areas and atolls, island shorelines, reef boundaries, and water bodies); MED (ports); and MLSA (atoll capital islands and cities).

Coastal Protection

Ocean waves, the changing mean sea level, beach erosion, and anthropogenic activities such as coral and sand mining threaten the coastal ecosystems. To protect their precious resources, the people of Maldives have identified sites to be safeguarded against coastal erosion. The coasts of 51 inhabited islands have coastal protection sites. Coastal protection is an important task for Maldivians as the coastal ecosystems—specifically the mangroves and coral reefs—also serve as a buffer from storm surges, inundation, and tsunamis. Having a well-protected coast also preserves the country's rich aquatic resources.

Table V.14: Maldives, Islands with Coastal Protection

Atoll	Number	Island	Atoll	Number	Island
Addu City	1	Feydhoo	Laamu	1	Gaadhoo
Alifu Alifu	3	Bodufolhudhoo	Lhaviyani	2	Hinnavaru
		Rasdhoo[a]			Kurehdhoo
		Ukulhas	Malé City	5	Hulhule
Alifu Dhaalu	2	Kun'burudhoo			HulhuMalé
		Maamigili			Malé[b]
Baa	3	Eydhafushi[a]			Thilafushi
		Fares			Vilin'gili
		Thulhaadhoo	Meemu	3	Dhiggaru
Dhaalu	4	Kudahuvadhoo[a]			Mulia
		Maaen'boodhoo			Naalaafushi
		Meedhoo	Noonu	2	Holhudhoo
		Rin'budhoo			Maafaru
Faafu	1	Nilandhoo[a]	Raa	3	Dhuvaafaru
Gaafu Alifu	3	Vilin'gili[a]			Fainu
		Faresmaathodaa			Maduvvari
		Gahdhoo	Shaviyani	2	Bileiyfahi
Haa Alifu	3	Dhidhdhoo[a]			Komandoo
		Ihavandhoo	Thaa	6	Dhiyamigili
		Kelaa			Guraidhoo

continued on next page

Table V.14 *continued*

Atoll	Number	Island	Atoll	Number	Island
Haa Dhaalu	3	Finey			Kan'doodhoo
		Kulhudhuffushi[a]			Madifushi
		Neykurendhoo			Thimarafushi
Kaafu	4	Guraidhoo			Vilufushi
		Himmafushi			
		Maafushi			
		Thulusdhoo[a]			

Notes:
[a] Atoll capital.
[b] City.

Source: Ministry of National Planning and Infrastructure, 2017.

Map V.43: Maldives, Coastal Protection

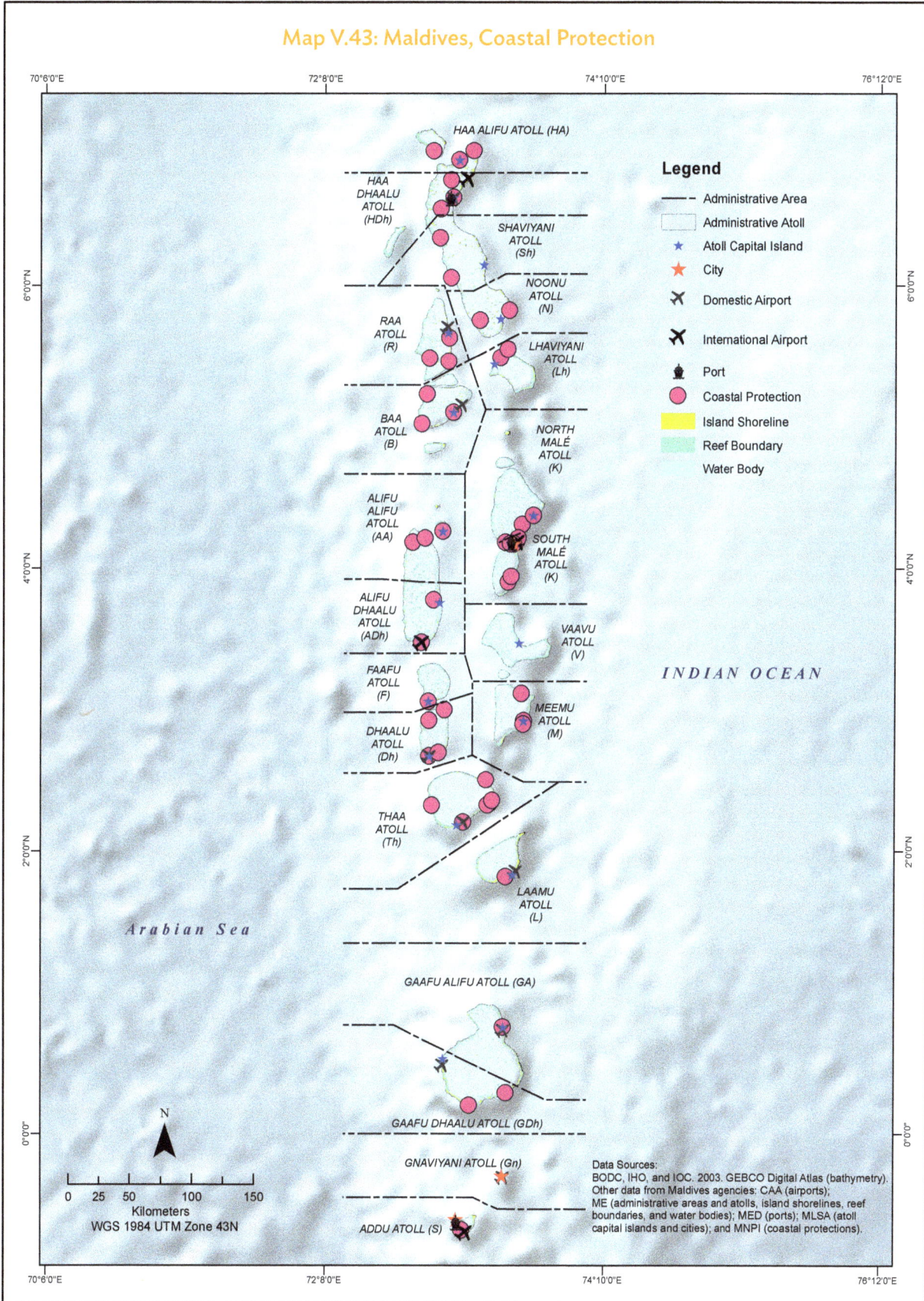

Legend
- — Administrative Area
- ☐ Administrative Atoll
- ★ Atoll Capital Island
- ★ City
- ✈ Domestic Airport
- ✈ International Airport
- ⚓ Port
- ● Coastal Protection
- ▭ Island Shoreline
- ▭ Reef Boundary
- Water Body

HAA ALIFU ATOLL (HA)
HAA DHAALU ATOLL (HDh)
SHAVIYANI ATOLL (Sh)
NOONU ATOLL (N)
RAA ATOLL (R)
LHAVIYANI ATOLL (Lh)
BAA ATOLL (B)
NORTH MALÉ ATOLL (K)
ALIFU ALIFU ATOLL (AA)
SOUTH MALÉ ATOLL (K)
ALIFU DHAALU ATOLL (ADh)
VAAVU ATOLL (V)
FAAFU ATOLL (F)
MEEMU ATOLL (M)
DHAALU ATOLL (Dh)
THAA ATOLL (Th)
LAAMU ATOLL (L)
GAAFU ALIFU ATOLL (GA)
GAAFU DHAALU ATOLL (GDh)
GNAVIYANI ATOLL (Gn)
ADDU ATOLL (S)

INDIAN OCEAN

Arabian Sea

N

Kilometers 0 25 50 100 150
WGS 1984 UTM Zone 43N

Data Sources:
BODC, IHO, and IOC. 2003. GEBCO Digital Atlas (bathymetry).
Other data from Maldives agencies: CAA (airports);
ME (administrative areas and atolls, island shorelines, reef
boundaries, and water bodies); MED (ports); MLSA (atoll
capital islands and cities); and MNPI (coastal protections).

Marine Conservation and Biodiversity

Maldives has greater coral reef coverage than available dry land. It has diverse and rich coral reefs, lagoons teeming with aquatic life, mangrove ecosystems where fingerlings as well as avian creatures find refuge, beaches where turtles lay their eggs, and seagrass beds. Human activities, coastal erosion, and changing climate patterns—specifically warmer seas—threaten the rich marine resources, particularly leading to coral bleaching. To protect this, the Marine Research Centre (now called 'Marine Research Institute') established coral reef monitoring sites in the 1990s to address the coral bleaching event that occurred in 1998 (Ibrahim et al. 2017). Coral reef monitoring sites can be found in the islands of Addu City and Alifu Alifu, Gaafu Alifu, Haa Dhaalu, North Malé, Seemu, and Vaavu atolls.

Table V.15: Maldives, Islands with Coral Reef Monitoring Sites

Atoll	Island
Haa Dhaalu	Hondaafushi
Haa Dhaalu	Finey
Haa Dhaalu	Hirimaradhoo
North Malé	Bodubandos
North Malé	Udhafushi
North Malé	Enboodhoofinolhu
Alifu Alifu	Fesdhoo
Alifu Alifu	Maayaafushi
Alifu Alifu	Velidhoo
Alifu Alifu	Kandhonlhudhoo
Vaavu	Anbaraa
Vaavu	Vattaru
Vaavu	Foththeyo
Gaafu Alifu	Kooddoo
Seemu	Hithadhoo
Seemu	Gan
Seemu	Vilingili

Source: Maldives Marine Research Centre, 2017.

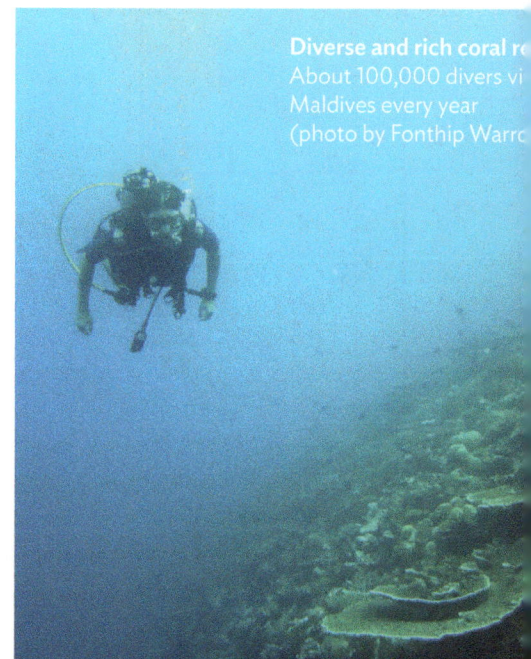

Diverse and rich coral re[...] About 100,000 divers vi[...] Maldives every year (photo by Fonthip Warro[...]

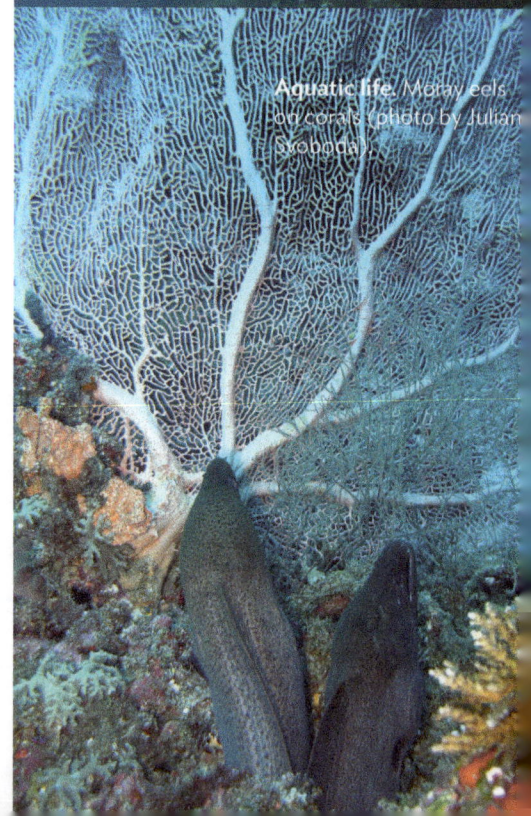

Aquatic life. Moray eels on corals (photo by Julian Svoboda)

Biodiversity. Coral reef teeming with life (photo by Fonthip Warrd).

Threats to Marine and Coastal Biodiversity

Coastal Erosion

The islands of Maldives, with an average elevation of 1.4 meters above sea level, are prone to coastal erosions and inundation. Natural factors such as tides, waves, and surges cause these coastal erosions. However, human activities such as sand mining increase the severity of beach erosion. The rising global mean sea level is another threatening factor. Global mean sea level, which is connected to rising temperature, would increase the country's coastal erosion. As of 2017, 45 islands are very severely eroded, 20 are severely eroded, and 18 are slightly eroded. Small island resorts are greatly vulnerable to coastal erosions and, as a result, are already losing economic gains (Emerton, Baig, and Saleem 2009).

Coastal erosion. Wave breakers are installed along the coast to protect the beach from erosion (photo by Erwin Österreich).

Threats to Marine and Coastal Biodiversity

Coral Bleaching

Maldivians benefit from rich coral reefs. From serving as home to aquatic resources—the nation's main source of food—to being one of the country's tourist attractions, coral reefs are undeniably essential to life in Maldives.

Protecting corals is integral to preserving Maldives' marine resources and keeping its fishing industry alive. While corals are now better monitored, threats such as rising temperatures still cannot be controlled.

In 2015–2016, the worst coral bleaching event in Maldives happened due to high temperatures associated with the El Niño phenomenon (Ibrahim et al. 2017). Based on climate projection, annual average temperature will increase by more than 1°C in 30 years (ADB 2017b). This could cause greater damage to corals in the coming years.

Coral bleaching. High seawater temperatures can cause coral bleaching, or the whitening of corals due to the loss of a symbiotic algae. Coral bleaching can lead to loss of individual corals and colonies, which in turn can cause the decline in population of numerous marine flora and fauna that depend on them.

Map V.44: Maldives, Coral Bleaching Risk Assessment

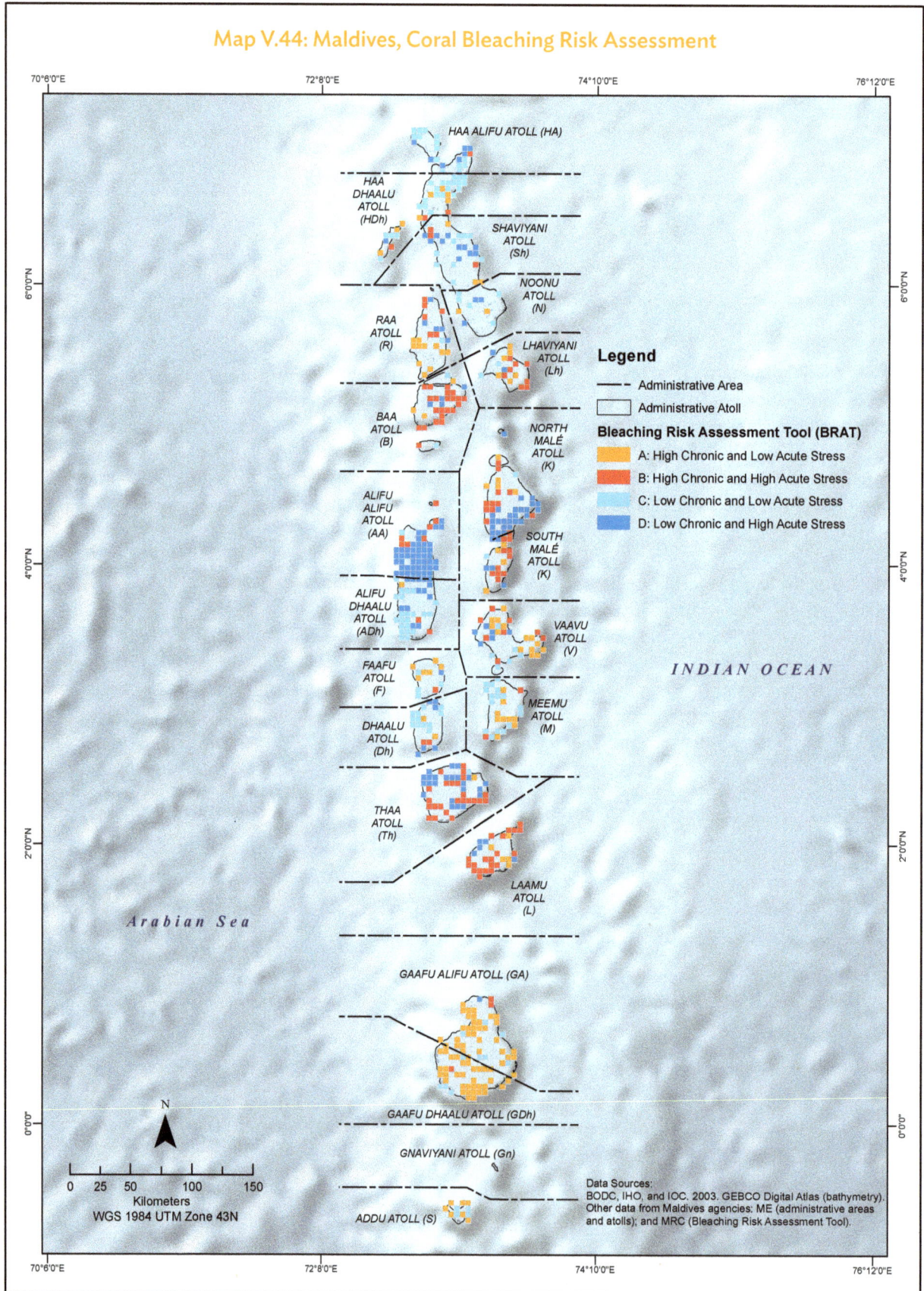

Legend

- – – Administrative Area
- ☐ Administrative Atoll

Bleaching Risk Assessment Tool (BRAT)

- A: High Chronic and Low Acute Stress
- B: High Chronic and High Acute Stress
- C: Low Chronic and Low Acute Stress
- D: Low Chronic and High Acute Stress

HAA ALIFU ATOLL (HA)
HAA DHAALU ATOLL (HDh)
SHAVIYANI ATOLL (Sh)
NOONU ATOLL (N)
RAA ATOLL (R)
LHAVIYANI ATOLL (Lh)
BAA ATOLL (B)
NORTH MALÉ ATOLL (K)
ALIFU ALIFU ATOLL (AA)
SOUTH MALÉ ATOLL (K)
ALIFU DHAALU ATOLL (ADh)
VAAVU ATOLL (V)
FAAFU ATOLL (F)
MEEMU ATOLL (M)
DHAALU ATOLL (Dh)
THAA ATOLL (Th)
LAAMU ATOLL (L)
GAAFU ALIFU ATOLL (GA)
GAAFU DHAALU ATOLL (GDh)
GNAVIYANI ATOLL (Gn)
ADDU ATOLL (S)

INDIAN OCEAN
Arabian Sea

0 25 50 100 150 Kilometers
WGS 1984 UTM Zone 43N

Data Sources:
BODC, IHO, and IOC. 2003. GEBCO Digital Atlas (bathymetry).
Other data from Maldives agencies: ME (administrative areas and atolls); and MRC (Bleaching Risk Assessment Tool).

Disaster Risk in Maldives

The four volumes of the *Multihazard Risk Atlas of Maldives* present the various components of disaster risk in the country. *Volume I* looks at the geography of Maldives, land reclamation, and land use and land cover in the islands. *Volume II* examines the historical and projected climate that could affect Maldivians as well as their natural flora and fauna, which are then mapped out in *Volumes III and IV*.

Humans, plants, animals, and physical structures for education, health, tourism, transportation, and power are all elements exposed to natural hazards. These hazards include climate, extreme weather, earthquakes, tsunamis, typhoons, surges, sea level rise, and others. The conditions of elements such as the presence of land reclamation, sand mining activities, and coastal erosion characterize the vulnerability of the exposed islands to storm surges, sea level rise, inundation, and tsunamis. Other factors, such as the human development index, power source, health, education, and transportation, define the vulnerability of the exposed population to various hazards. Environmentally sensitive areas and bleached corals indicate an increased vulnerability of the ecosystem to environmental stresses and hazards, while having coastal protection and monitoring sites indicates adaptation capacity, which lowers vulnerability to disasters.

The following maps were prepared based on indexed risk tables, which were generated from detailed analysis of relevant data obtained from ME and the National Disaster Management Center of Maldives. These maps include physical and social risks. In addition, physical vulnerability or susceptibility maps feature multihazard hydrometeorological as well as rain-induced flooding; tsunamis; big waves or *udha*; and wave, rain, and wind hazards. Table V.16 shows the numerical ranges of the hazard categories mapped.

Table V.16: Hazard Categories and Index Ranges

Category	Minimum	Maximum
Rain-induced Flood		
Low	0.00	0.10
Medium	0.11	0.25
High	0.45	1.26
Udha		
Low	0.00	0.07
Medium	0.14	0.32
High	0.71	7.65
Wave, Rain, Wind (Flood) Hazard		
Low	0.00	0.05
Medium	0.10	0.33
High	0.43	1.43
Wind and Wave Hazard		
Low	0.00	0.09
Medium	0.10	0.36
High	0.45	5.26
Hydrometeorological Multihazard		
Low	0.009	
Medium	0.132	0.367
High	0.403	15.301

udha = big wave.

Source: Asian Development Bank.

Map V.45: Maldives, Rain-Induced Flood (Islands)

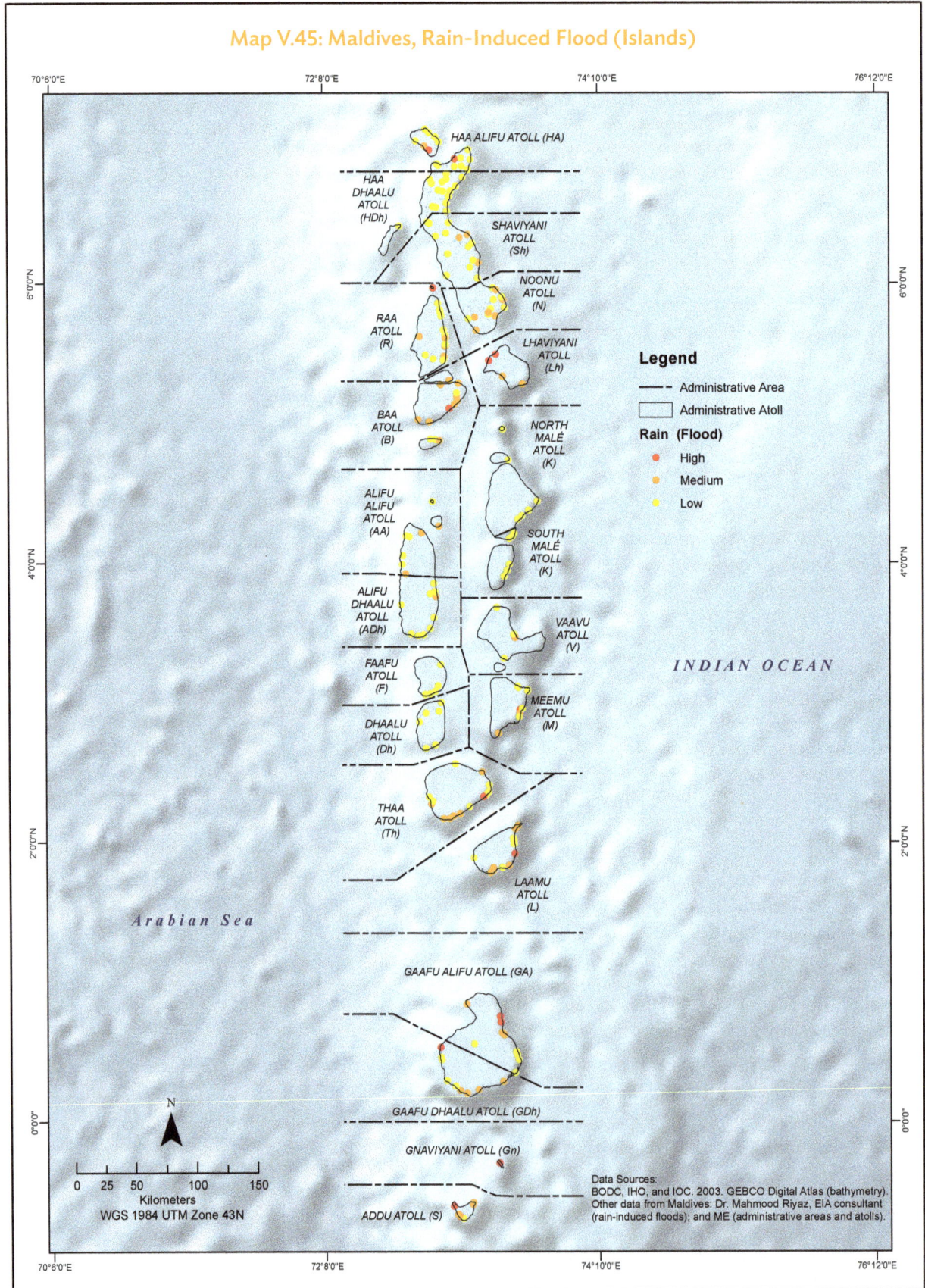

HAA ALIFU ATOLL (HA)

HAA DHAALU ATOLL (HDh)

SHAVIYANI ATOLL (Sh)

NOONU ATOLL (N)

RAA ATOLL (R)

LHAVIYANI ATOLL (Lh)

BAA ATOLL (B)

NORTH MALÉ ATOLL (K)

ALIFU ALIFU ATOLL (AA)

SOUTH MALÉ ATOLL (K)

ALIFU DHAALU ATOLL (ADh)

VAAVU ATOLL (V)

FAAFU ATOLL (F)

MEEMU ATOLL (M)

DHAALU ATOLL (Dh)

THAA ATOLL (Th)

LAAMU ATOLL (L)

GAAFU ALIFU ATOLL (GA)

GAAFU DHAALU ATOLL (GDh)

GNAVIYANI ATOLL (Gn)

ADDU ATOLL (S)

INDIAN OCEAN

Arabian Sea

Legend

— · — Administrative Area

▭ Administrative Atoll

Rain (Flood)

● High
● Medium
● Low

N

0 25 50   100   150
Kilometers
WGS 1984 UTM Zone 43N

Data Sources:
BODC, IHO, and IOC. 2003. GEBCO Digital Atlas (bathymetry).
Other data from Maldives: Dr. Mahmood Riyaz, EIA consultant
(rain-induced floods); and ME (administrative areas and atolls).

Map V.46: Maldives, *Udha* Hazard (Islands)

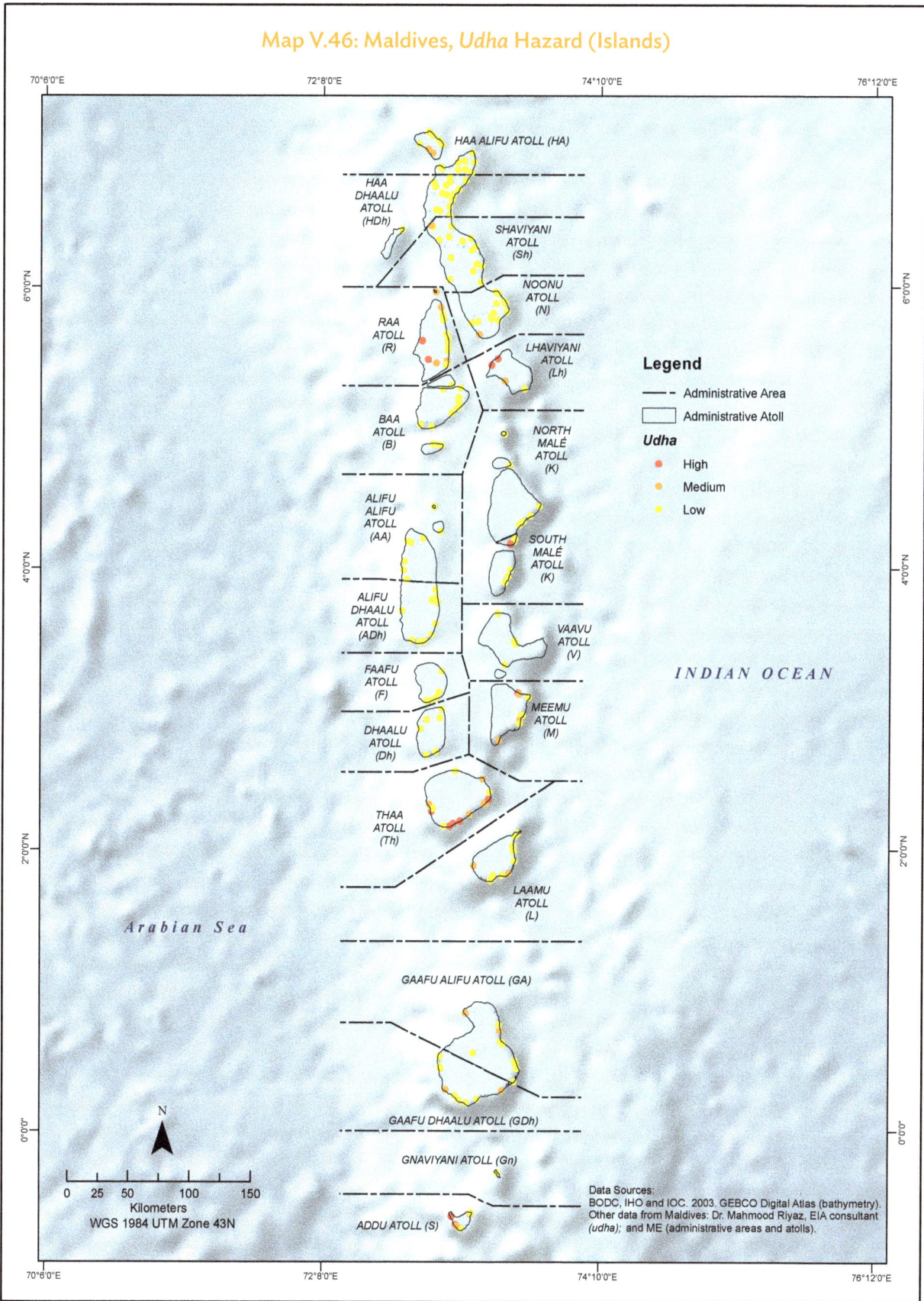

HAA ALIFU ATOLL (HA)

HAA DHAALU ATOLL (HDh)

SHAVIYANI ATOLL (Sh)

NOONU ATOLL (N)

RAA ATOLL (R)

LHAVIYANI ATOLL (Lh)

BAA ATOLL (B)

NORTH MALÉ ATOLL (K)

ALIFU ALIFU ATOLL (AA)

SOUTH MALÉ ATOLL (K)

ALIFU DHAALU ATOLL (ADh)

VAAVU ATOLL (V)

FAAFU ATOLL (F)

MEEMU ATOLL (M)

DHAALU ATOLL (Dh)

THAA ATOLL (Th)

LAAMU ATOLL (L)

INDIAN OCEAN

Arabian Sea

GAAFU ALIFU ATOLL (GA)

GAAFU DHAALU ATOLL (GDh)

GNAVIYANI ATOLL (Gn)

ADDU ATOLL (S)

Legend

— · — Administrative Area

▭ Administrative Atoll

Udha

● High
● Medium
● Low

N

0 25 50 100 150
Kilometers
WGS 1984 UTM Zone 43N

Data Sources:
BODC, IHO and IOC. 2003. GEBCO Digital Atlas (bathymetry).
Other data from Maldives: Dr. Mahmood Riyaz, EIA consultant
(udha); and ME (administrative areas and atolls).

Map V.47: Maldives, Wind and Wave Hazards

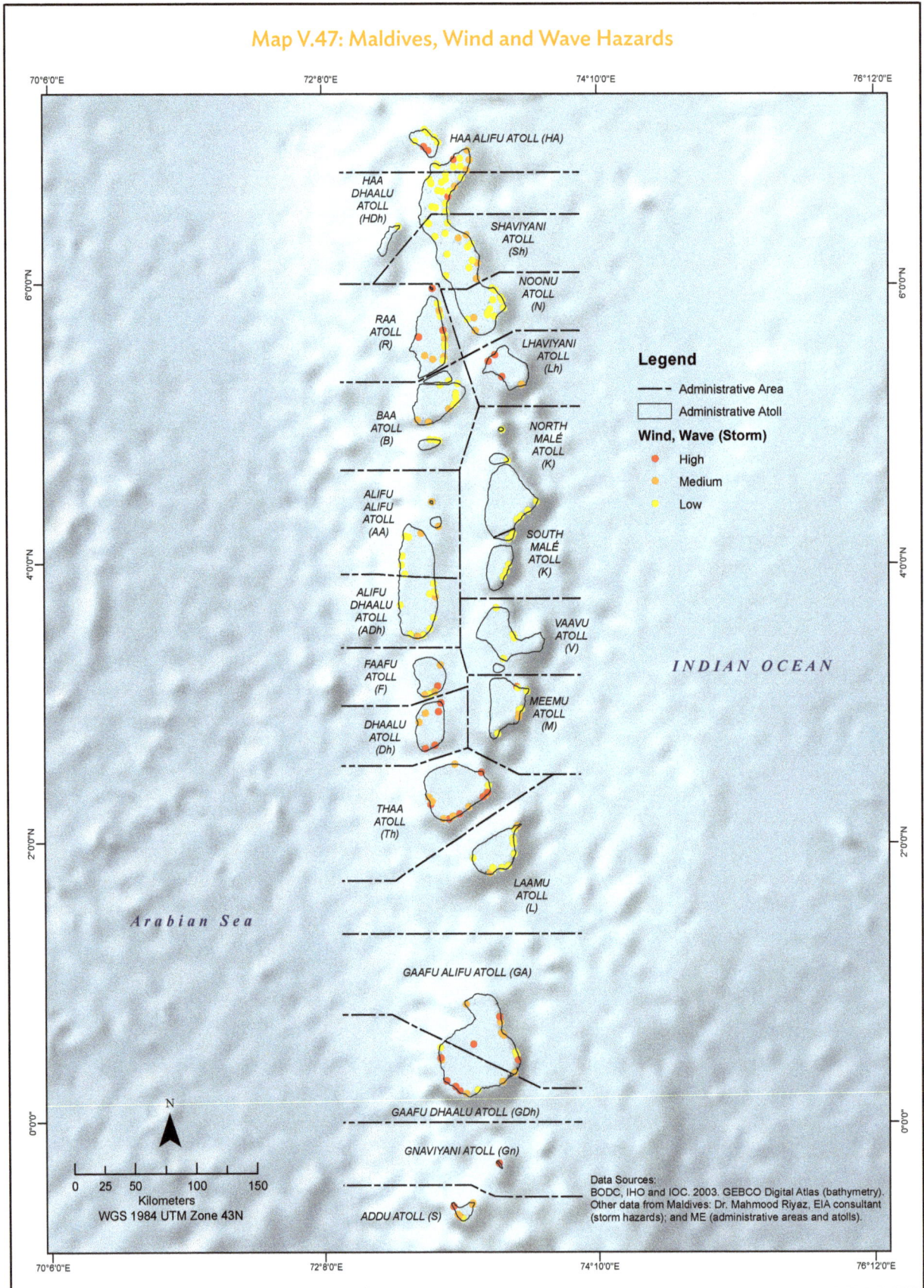

HAA ALIFU ATOLL (HA)

HAA DHAALU ATOLL (HDh)

SHAVIYANI ATOLL (Sh)

NOONU ATOLL (N)

RAA ATOLL (R)

LHAVIYANI ATOLL (Lh)

BAA ATOLL (B)

NORTH MALÉ ATOLL (K)

ALIFU ALIFU ATOLL (AA)

SOUTH MALÉ ATOLL (K)

ALIFU DHAALU ATOLL (ADh)

VAAVU ATOLL (V)

FAAFU ATOLL (F)

MEEMU ATOLL (M)

DHAALU ATOLL (Dh)

THAA ATOLL (Th)

LAAMU ATOLL (L)

GAAFU ALIFU ATOLL (GA)

GAAFU DHAALU ATOLL (GDh)

GNAVIYANI ATOLL (Gn)

ADDU ATOLL (S)

INDIAN OCEAN

Arabian Sea

Legend

– – – Administrative Area

☐ Administrative Atoll

Wind, Wave (Storm)

● High

● Medium

● Low

N

0   25   50   100   150
Kilometers
WGS 1984 UTM Zone 43N

Data Sources:
BODC, IHO and IOC. 2003. GEBCO Digital Atlas (bathymetry).
Other data from Maldives: Dr. Mahmood Riyaz, EIA consultant
(storm hazards); and ME (administrative areas and atolls).

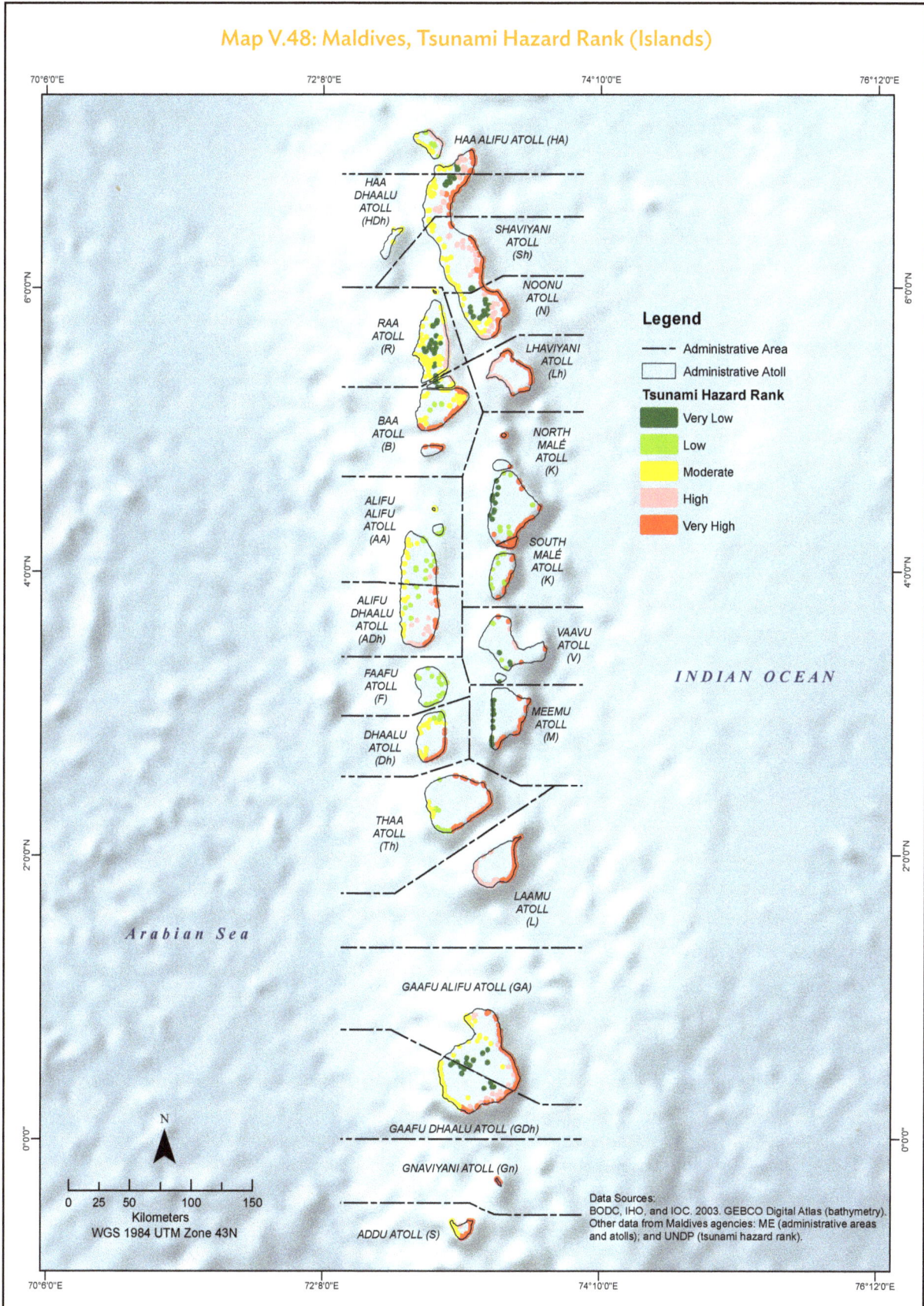

Map V.48: Maldives, Tsunami Hazard Rank (Islands)

Map V.49: Maldives, Hydrometeorological Multihazard (Islands)

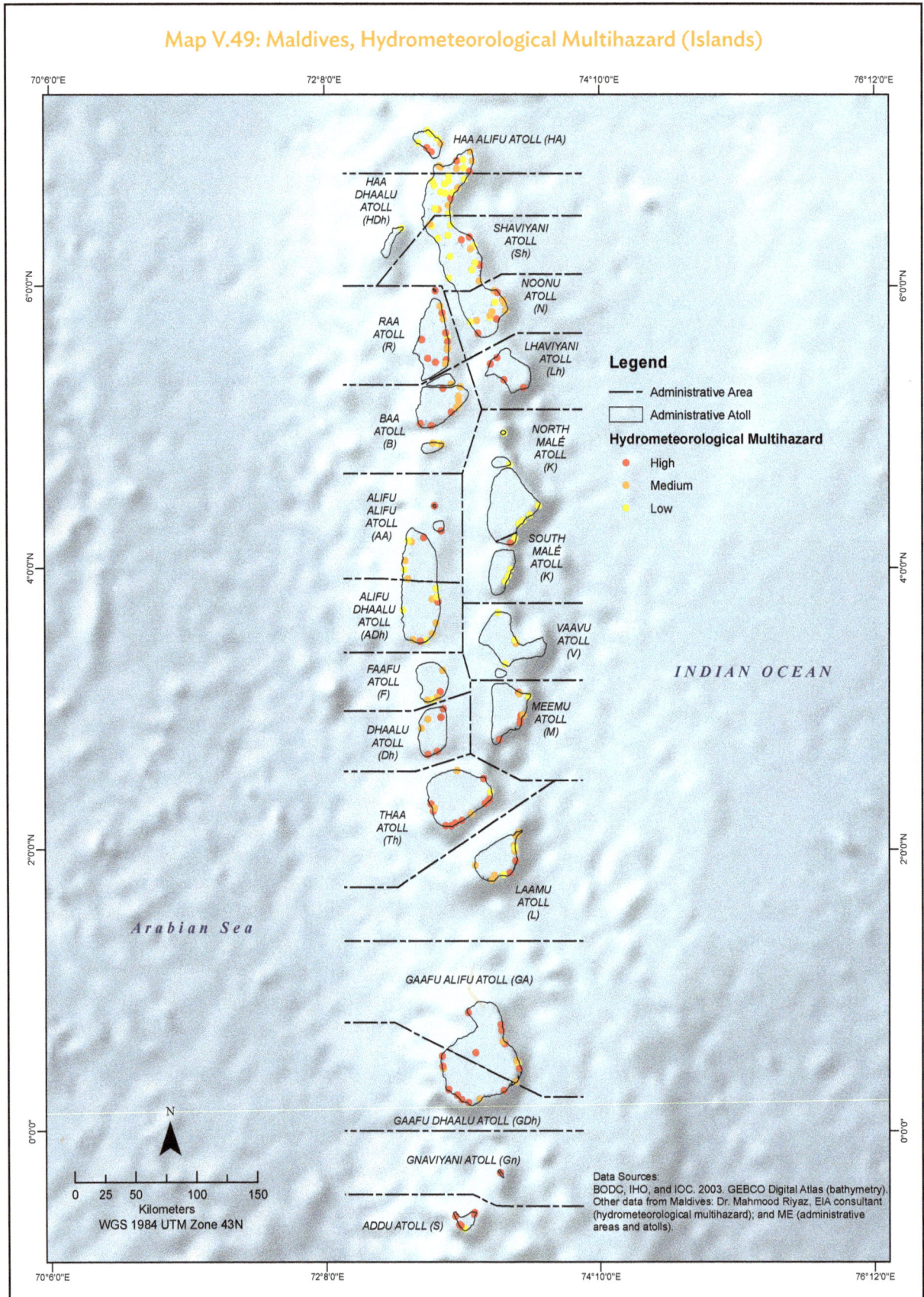

HAA ALIFU ATOLL (HA)

HAA DHAALU ATOLL (HDh)

SHAVIYANI ATOLL (Sh)

NOONU ATOLL (N)

RAA ATOLL (R)

LHAVIYANI ATOLL (Lh)

BAA ATOLL (B)

NORTH MALÉ ATOLL (K)

ALIFU ALIFU ATOLL (AA)

SOUTH MALÉ ATOLL (K)

ALIFU DHAALU ATOLL (ADh)

VAAVU ATOLL (V)

FAAFU ATOLL (F)

MEEMU ATOLL (M)

DHAALU ATOLL (Dh)

THAA ATOLL (Th)

LAAMU ATOLL (L)

GAAFU ALIFU ATOLL (GA)

GAAFU DHAALU ATOLL (GDh)

GNAVIYANI ATOLL (Gn)

ADDU ATOLL (S)

INDIAN OCEAN

Arabian Sea

Legend

— · — Administrative Area

☐ Administrative Atoll

Hydrometeorological Multihazard

● High

● Medium

● Low

N

0 25 50   100   150
Kilometers
WGS 1984 UTM Zone 43N

Data Sources:
BODC, IHO, and IOC. 2003. GEBCO Digital Atlas (bathymetry).
Other data from Maldives: Dr. Mahmood Riyaz, EIA consultant
(hydrometeorological multihazard); and ME (administrative
areas and atolls).

Map V.50: Maldives, Multihazard Physical Risk Index

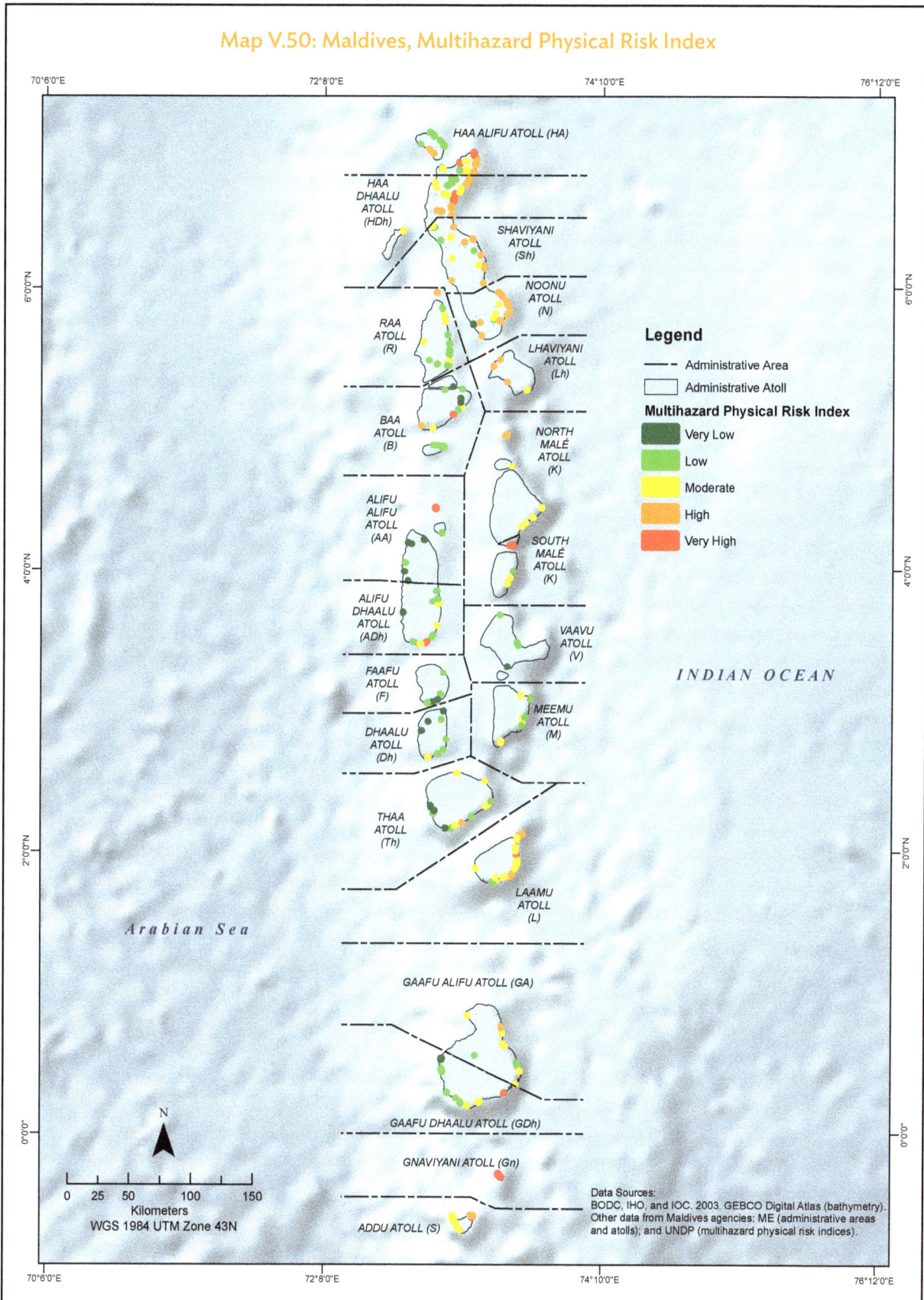

Legend

– – – Administrative Area

☐ Administrative Atoll

Multihazard Physical Risk Index

- Very Low
- Low
- Moderate
- High
- Very High

INDIAN OCEAN

Arabian Sea

HAA ALIFU ATOLL (HA)
HAA DHAALU ATOLL (HDh)
SHAVIYANI ATOLL (Sh)
NOONU ATOLL (N)
RAA ATOLL (R)
LHAVIYANI ATOLL (Lh)
BAA ATOLL (B)
NORTH MALÉ ATOLL (K)
ALIFU ALIFU ATOLL (AA)
SOUTH MALÉ ATOLL (K)
ALIFU DHAALU ATOLL (ADh)
VAAVU ATOLL (V)
FAAFU ATOLL (F)
MEEMU ATOLL (M)
DHAALU ATOLL (Dh)
THAA ATOLL (Th)
LAAMU ATOLL (L)
GAAFU ALIFU ATOLL (GA)
GAAFU DHAALU ATOLL (GDh)
GNAVIYANI ATOLL (Gn)
ADDU ATOLL (S)

N

0 25 50 100 150
Kilometers
WGS 1984 UTM Zone 43N

Data Sources:
BODC, IHO, and IOC. 2003. GEBCO Digital Atlas (bathymetry).
Other data from Maldives agencies: ME (administrative areas
and atolls); and UNDP (multihazard physical risk indices).

Map V.51: Maldives, Multihazard Social Risk Index

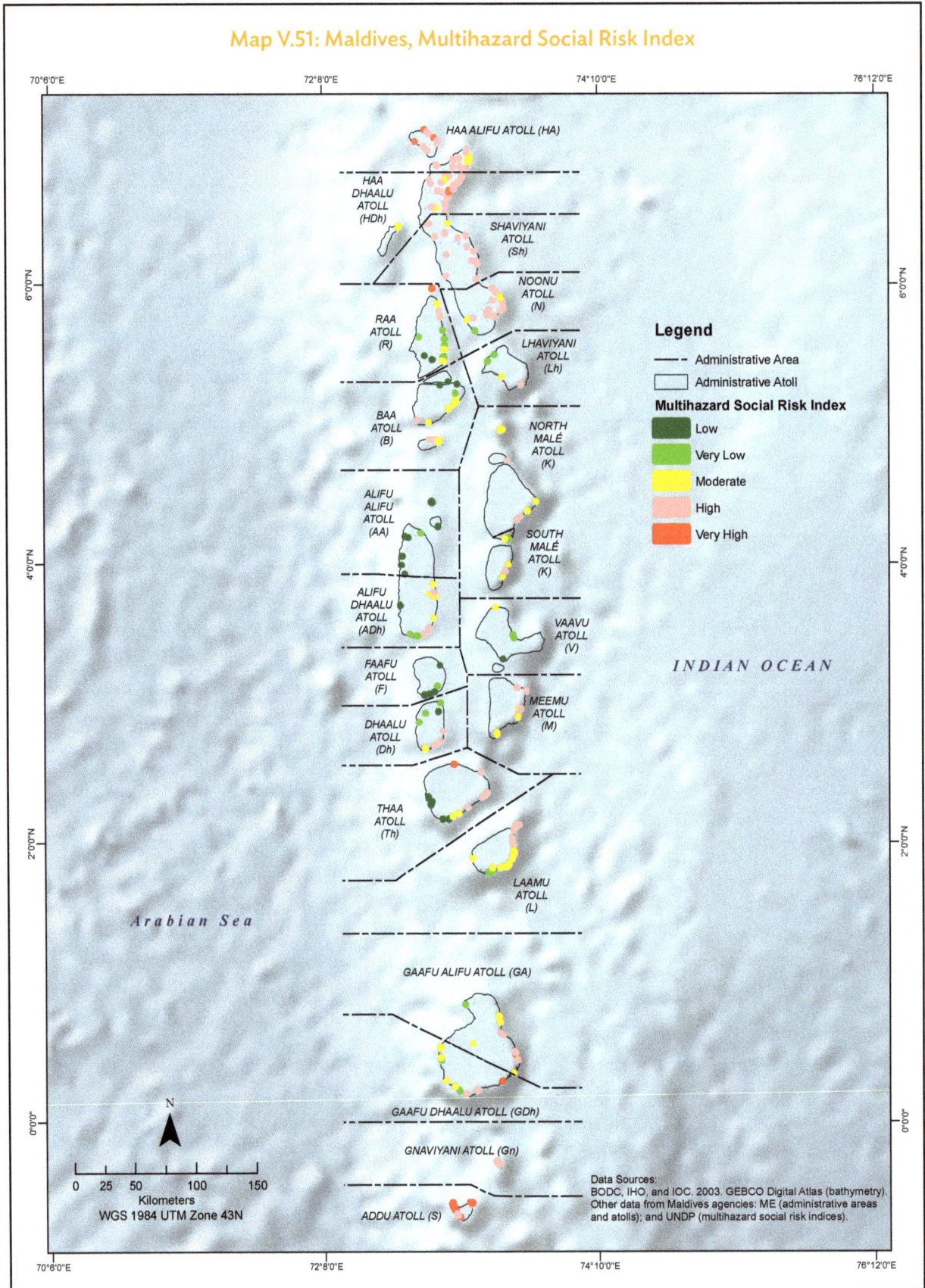

Legend

- — · — Administrative Area
- ☐ Administrative Atoll

Multihazard Social Risk Index

- Low
- Very Low
- Moderate
- High
- Very High

HAA ALIFU ATOLL (HA)

HAA DHAALU ATOLL (HDh)

SHAVIYANI ATOLL (Sh)

NOONU ATOLL (N)

RAA ATOLL (R)

LHAVIYANI ATOLL (Lh)

BAA ATOLL (B)

NORTH MALÉ ATOLL (K)

ALIFU ALIFU ATOLL (AA)

SOUTH MALÉ ATOLL (K)

ALIFU DHAALU ATOLL (ADh)

VAAVU ATOLL (V)

FAAFU ATOLL (F)

MEEMU ATOLL (M)

DHAALU ATOLL (Dh)

THAA ATOLL (Th)

LAAMU ATOLL (L)

GAAFU ALIFU ATOLL (GA)

GAAFU DHAALU ATOLL (GDh)

GNAVIYANI ATOLL (Gn)

ADDU ATOLL (S)

INDIAN OCEAN

Arabian Sea

N

0 25 50 100 150
Kilometers
WGS 1984 UTM Zone 43N

Data Sources:
BODC, IHO, and IOC. 2003. GEBCO Digital Atlas (bathymetry).
Other data from Maldives agencies: ME (administrative areas
and atolls); and UNDP (multihazard social risk indices).

Map Data Sources

Government Ministries, Departments, and Agencies in Maldives
    Civil Aviation Authority
        Airports
        Floatplane platform
    Environmental Protection Agency
        Coastal erosion
        Environmentally protected and sensitive areas
    Land and Survey Authority
        Atoll capital islands
        Cities
    Marine Research Institute
        Coral reef monitoring sites
    Meteorological Service
        Automatic weather stations
        Meteorological observation stations
    Ministry of Economic Development
        Ports
    Ministry of Education
        Education centers
    Ministry of Environment
        Administrative areas
        Administrative atolls
        Island shorelines
        Land use and land cover
        Mangroves
        Power stations
        Reef boundaries
        Water bodies
    Ministry of Fisheries, Marine Resources and Agriculture
        Fish aggregating devices
    Ministry of Health
        Healthcare facilities
    Ministry of National Planning and Infrastructure
        Coastal protection
        Harbor facilities
        Land reclamation

Ministry of Tourism
    Resort islands
National Bureau of Statistics
    Population

International Institutions
    Marine Spatial Ecology Lab, University of Queensland, Australia
        Bleaching Risk Assessment Tool
    SANDER + PARTNER. www.Sander-Partner.com
        Rainfall
        Temperature
    United Nations Development Programme
        Cyclonic wind hazard zone
        Human development index 2014
        Multihazard physical risk index for the island
        Multihazard social risk index for the island
        Seismic hazard zone
        Surge hazard zone
        Tsunami hazard rank

International Institution in Maldives
    International Union for Conservation of Nature, Maldives
        Approved sand mining locations

Private Individual
    Mahmood Riyaz, EPA-Licensed Environmental Impact Assessment Specialist
        Hydrometeorological multihazard
        Rain-induced flood
        Udha vulnerability
        Wave, rain, wind (flood) vulnerability
        Wind, wave (storm) vulnerability

References

Ahmed, M., and S. Suphachalasai. 2014. *Assessing the Costs of Climate Change and Adaptation in South Asia.* Manila: Asian Development Bank, UK Aid.

Asian Development Bank. 2015. *Maldives: Overcoming the Challenges of a Small Island State. Country Diagnostic Study.* Manila.

ADB. 2017a. *Climate Risk Screening for Mainstreaming Climate Change Adaptation into Development Activities and Policies in the Maldives.* Consultants' report. Manila (TA 8572-REG).

ADB. 2017b. *Final Report on Climate Downscaling for Bangladesh, Maldives, and Sri Lanka.* Consultants' report. Manila (TA 8572-REG).

British Oceanographic Data Centre (BODC), International Hydrographic Organisation (IHO) and the Intergovernmental Oceanographic Commission (IOC) of the United Nations Educational, Scientific and Cultural Organization. 2003. *General Bathymetric Chart of the Oceans (GEBCO) Digital Atlas.* UK: British Oceanographic Data Centre.

Dhunya, A., Q. Huang, and A. Aslam. 2017. Coastal Habitats of Maldives: Status, Trends, Threats, and Potential Conservation Strategies. *International Journal of Scientific and Engineering Research.* 8 (3). pp. 47–49.

Emerton, L., S. Baig, and M. Saleem. 2009. *Valuing Biodiversity: The Economic Case for Biodiversity Conservation in the Maldives.* Homagama: Ecosystems and Livelihoods Group Asia; International Union for the Conservation of Nature for the Atoll Ecosystem Conservation Project; Ministry of Housing, Transport, and Environment, Government of Maldives.

Hosterman, H., and J. Smith. 2014. *Economic Costs and Benefits of Climate Change Impacts and Adaptation to the Maldives Tourism Industry.* Malé: Ministry of Tourism.

Ibrahim, N., M. Mohamed, A. Basheer, H. Ismail, F. Nistharan, A. Schmidt, R. Naeem, A. Abdulla, and G. Grimsditch. 2017. *Status of Coral Bleaching in the Maldives in 2016.* Malé: Marine Research Centre.

Intergovernmental Panel on Climate Change. 2001. *Climate Change 2001: The Scientific Basis. Contribution of Working Group I to the Third Assessment Report of the Intergovernmental Panel on Climate Change.* Cambridge, United Kingdom, and New York, USA: Cambridge University Press.

Intergovernmental Panel on Climate Change. 2014. *Climate Change 2014: Synthesis Report. Contribution of Working Groups I, II and III to the Fifth Assessment Report of the Intergovernmental Panel on Climate Change* [Core Writing Team, R.K. Pachauri, and L.A. Meyer, eds.]. Geneva: IPCC.

Khan, T., D. Quadir, T. Murty, A. Kabir, F. Aktar, and M. Sarker. 2002. Relative Sea Level Changes in Maldives and Vulnerability of Land Due to Abnormal Coastal Inundation. *Marine Geodesy.* 25 (1–2). pp. 133–143.

Ministry of Environment and Energy. 2016. *State of the Environment.* Malé: Ministry of Environment and Energy.

Moosa, F. S. 2014. *Country Report: Republic of Maldives.* Asian Disaster Reduction Centre.

Naseer, A. 1997. *Paper 5: Status of Coral Mining in the Maldives: Impacts and Management Options.* Food and Agriculture Organization of the United Nations. http://www.fao.org/docrep/X5623E/x5623e0o.htm.

National Bureau of Statistics. 2014. *Maldives Population and Housing Census: Statistical Release 1 - Population and Households.* Malé.

United Nations Development Programme. 2006. *Developing a Disaster Risk Profile for Maldives.*

United Nations Development Programme. 2018. *Briefing Note for Countries on the 2018 Statistical Update: Maldives.* http://hdr.undp.org/sites/all/themes/hdr_theme/country-notes/MDV.pdf.

United Nations Educational, Scientific and Cultural Organization's Institute for Statistics. 2014. *Maldives.* https://en.unesco.org/countries/maldives.

Waheed, M., and H. Shakoor. 2015. The Impact of the Indian Ocean Tsunami on Maldives. In C. Bassard, D. W. Giles, and A. M. Howitt, eds. *Natural Disaster Management in the Asia-Pacific: Policy and Governance.* New York: Springer. pp. 49–68.

Maps were prepared by the Country Consultant Team and the Manila Observatory on behalf of the Asian Development Bank.